AF667471

Metals in Soil - Contamination and Remediation

Edited by Zinnat Ara Begum, Ismail M. M. Rahman and Hiroshi Hasegawa

Published in London, United Kingdom

IntechOpen

Supporting open minds since 2005

Metals in Soil - Contamination and Remediation
http://dx.doi.org/10.5772/intechopen.77986
Edited by Zinnat Ara Begum, Ismail M. M. Rahman and Hiroshi Hasegawa

Contributors
Fidèle Suanon, Alexis Finagnon Crepin Togbe, Lyde A. S. Tomètin, Binessi Edouard Ifon, Arouan Yessoufou, Muhammad Wasim, Shahrukh Shoaib, Om Bansal, Jelena Markovic, Svetlana Stevovic, Zinnat Ara Begum

© The Editor(s) and the Author(s) 2019
The rights of the editor(s) and the author(s) have been asserted in accordance with the Copyright, Designs and Patents Act 1988. All rights to the book as a whole are reserved by INTECHOPEN LIMITED. The book as a whole (compilation) cannot be reproduced, distributed or used for commercial or non-commercial purposes without INTECHOPEN LIMITED's written permission. Enquiries concerning the use of the book should be directed to INTECHOPEN LIMITED rights and permissions department (permissions@intechopen.com).
Violations are liable to prosecution under the governing Copyright Law.

Individual chapters of this publication are distributed under the terms of the Creative Commons Attribution 3.0 Unported License which permits commercial use, distribution and reproduction of the individual chapters, provided the original author(s) and source publication are appropriately acknowledged. If so indicated, certain images may not be included under the Creative Commons license. In such cases users will need to obtain permission from the license holder to reproduce the material. More details and guidelines concerning content reuse and adaptation can be found at http://www.intechopen.com/copyright-policy.html.

Notice
Statements and opinions expressed in the chapters are these of the individual contributors and not necessarily those of the editors or publisher. No responsibility is accepted for the accuracy of information contained in the published chapters. The publisher assumes no responsibility for any damage or injury to persons or property arising out of the use of any materials, instructions, methods or ideas contained in the book.

First published in London, United Kingdom, 2019 by IntechOpen
IntechOpen is the global imprint of INTECHOPEN LIMITED, registered in England and Wales, registration number: 11086078, The Shard, 25th floor, 32 London Bridge Street
London, SE19SG - United Kingdom
Printed in Croatia

British Library Cataloguing-in-Publication Data
A catalogue record for this book is available from the British Library

Additional hard copies can be obtained from orders@intechopen.com

Metals in Soil - Contamination and Remediation
Edited by Zinnat Ara Begum, Ismail M. M. Rahman and Hiroshi Hasegawa
p. cm.
Print ISBN 978-1-78985-775-7
Online ISBN 978-1-78985-776-4

For EU product safety concerns:
IN TECH d.o.o., Prolaz Marije Krucifikse Kozulić 3, 51000 Rijeka, Croatia, info@intechopen.com or visit our website at intechopen.com.

We are IntechOpen,
the world's leading publisher of
Open Access books
Built by scientists, for scientists

4,000+
Open access books available

116,000+
International authors and editors

120M+
Downloads

151
Countries delivered to

Our authors are among the
Top 1%
most cited scientists

12.2%
Contributors from top 500 universities

WEB OF SCIENCE™

Selection of our books indexed in the Book Citation Index
in Web of Science™ Core Collection (BKCI)

Interested in publishing with us?
Contact book.department@intechopen.com

Numbers displayed above are based on latest data collected.
For more information visit www.intechopen.com

Meet the editors

Zinnat Ara Begum received her PhD degree in 2012 from Kanazawa University, Japan. She is currently working as a Research Fellow in the Venture Business Laboratory, Organization of Frontier Science and Innovation, Kanazawa University, Japan. She is also affiliated to the Department of Civil Engineering, Southern University, Chittagong, Bangladesh, as an Assistant Professor.

Ismail M. M. Rahman received his PhD degree in 2011 from Kanazawa University, Japan. He is currently working as an Associate Professor in the Institute of Environmental Radioactivity, Fukushima University, Japan.

Hiroshi Hasegawa received his DSc degree in 1997 from Kyoto University, Japan. He is currently working as a Professor in the Faculty of Chemistry, Institute of Science and Engineering, Kanazawa University, Japan.

Contents

Preface

The distribution and abundance of metals in soil changes from the corresponding natural ratio, presumably due to anthropogenic inputs. The intrusion of metal in soil induces several physiological and biochemical changes, which evoke ecological concerns. This book consists of reviews and case studies from different researchers focusing on different aspects of soil contamination by metals and their subsequent remediation. All the authors were invited by the publisher, who also declared accountability for the accuracy of their contributions.

The first section of the book (Metals in Soil—Contamination) starts with an introductory chapter that discusses the assessment of metal contamination in soil. The following chapter provides an overview of the effects of metal intrusion on the natural properties of soils, and the two following chapters in this section describe case studies related to anthropogenic impacts of metal accumulation in soil. The second section (Metals in Soil—Remediation) contains a single chapter describing the remediation options to treat metal-contaminated soil.

This multi-authored book is expected to provide a complete package of facts and issues related to metals in soil. We appreciate the efforts from the authors for their contributions and wish to thank InTechOpen for giving us the opportunity to serve as editors. The book would not have been possible without the sincere support from Author Service Manager Ms. Maja Bozicevic.

Zinnat Ara Begum
Kanazawa University, Japan,
Southern University Bangladesh, Bangladesh

Ismail M. M. Rahman
Fukushima University, Japan

Hiroshi Hasegawa
Kanazawa University, Japan

Section 1

Metals in Soil - Contamination

Chapter 1

Introductory Chapter: How to Assess Metal Contamination in Soils?

Ismail M.M. Rahman and Zinnat A. Begum

1. Introduction

The average concentrations of metal or metalloid referred as metal(loid) hereafter, except those of radioisotopes or daughter nuclides and inert gases, have remained virtually unchanged in the earth's crust despite the ups and downs in the overall distribution [1]. The total element content in the earth's crust is dominated by O, Si, Al, Fe, Ca, Na, K, Mg, P, and Ti representing ≥99%, while the other elements in the periodic table comprised the remaining 1% and are termed as "trace elements" [2]. The abundances of naturally occurring metal(loid)s in the earth's crust, also known as *Clarke values*, have been estimated by several researchers [3–5]. The *Clarke values* in different reports slightly varied because these are hypothetical concentrations as computed using assumed proportions of various crustal rock types [6]. The ore minerals, which contain significant contents of several metal(loid)s in their crystal structure, are listed in **Table 1**.

The changes in both distribution and abundances of metal(loid)s in the ecosphere have become catastrophically high in recent decades presumably attributable to a wide range of anthropogenic inputs [2]. The anthropogenic emission of the toxic metal(loid)s into the atmosphere is estimated to be the one-to-three order of magnitude higher than the natural fluxes [7]. Soil, an ecosphere compartment, is the primary sink for metal(loid)s released into the environment by anthropogenic activities, which often persist for an indefinite period as most metal(loid)s resist the microbial or chemical degradation [8, 9]. Metal(loid)s are usually adsorbed by the organic, inorganic, or colloidal constituents of soil, e.g., humus, hydrous oxides, and hydroxides of Al, Fe, or Mn and Al, phyllosilicates, and some sparingly soluble calcium salts [10]. However, the anthropogenic contaminants such as ash, mine waste, demolition rubble, and so forth can serve as the parent material of a nonnatural soil type, namely, Anthrosols [2], which should have different metal accumulation characteristics than the natural pedogenic soils. The anthropogenic metal(loid)s in soils might have increased mobility than those from pedogenic or genic origins [11]. The metal(loid) contamination of soil is colorless, odorless, and barely noticeable as the environmental impact is not expeditious. The ecological damage due to the metal(loid)s triggered when the corresponding bioavailability is above the threshold or there is a change of environmental conditions [12, 13]. Moreover, the impact of contamination is enhanced when multiple metal(loid)s are involved rather than a single species [14]. The magnitude of metal(loid)s concentration in soils depends on the type of exposure and may be varied on different sites. The physicochemical characteristics and the distribution of metal(loid)s diversified based on the interaction with the soils and local transport mechanisms [15, 16]. The adverse effects on soils due to the accumulation of metal(loid)s are summarized in **Table 2**.

IntechOpen

The environmental and geochemical changes of soils as a result of the intrusion of metal(loid)s not only affect the safety of living beings but also hamper the sustainable development due to the impact on the economic or political considerations

Ore minerals	Associated metalloids
Argentite (Ag_2S), PbS	Ag, Au, Cu, Sb, Zn, Pb, Se, Te
Arsenopyrite (FeAsS), AsS	As, Au, Ag, Sb, Hg, U, Bi, Mo, Sn, Cu
Barite ($BaSO_4$)	Ba, Pb, Zn
Sphalerite (ZnS)	Cd, Zn, Pb, Cu
Cobaltite ((Co, Fe) AsS)	Co, Fe, As, Sb, Cu, Ni, Ag, U
Chromite (Fe, Cr_2O_4)	Cr, Ni, Co
Bornite (Cu_5FeS_4), chalcocite (Cu_2S), chalcopyrite ($CuFeS_2$)	Cu, Zn, Pb, Cd, As, Se, Sb, Ni, Pt, Mo, Au, Te
Cinnabar (HgS)	Hg, Sb, Se, Te, Ag, Zn, Pb, Mn
Pyrolusite (MnO_2)	Mn, Co, Ni, Zn, Pb
Molybdenite (MoS_2)	Mo, Cu, Re, W, Sn
Galena (PbS)	Pb, Ag, Zn, Cu, Cd, Sb, Tl, Se, Te
Stibnite (Sb_2S_3)	Sb, Ag, Au, Hg, As
Cassiterite (SnO_2)	Sn, Nb, Ta, W, Rb
Uraninite (UO_2)	U, V, As, Mo, Se, Pb, Cu, Co,
Vanadinite ($Pb_5(VO_4)_3Cl$)	V, U, Pb
Wolframite ((Fe, Mn) WO_4)	W, Mo, Sn, Nb
Sphalerite (ZnS), smithsonite ($ZnCO_3$)	Zn, Cd, Cu, Pb, As, Se, Sb, Ag, In

†*Source: Alloway [2].*

Table 1.
Common source of ore minerals of the metal(loid)s.†

Agricultural effect	Reduction of soil fertility Reduction of nitrogen fixation Increased erosion factor Increasing soil loss Increase nutrient deficiency Reduction of crop yields Imbalance in the soil biota (flora, fauna, microorganism) Decrease of soil biodiversity
Industrial effect	Transfer of dangerous chemicals Ecological imbalance Release of pollutant gases Increased salinity
Urban effect	Clogging of the drains Soil deposits Flooding areas Health problems Contamination of drinking water sources Problems of waste management

†*Source: Weissmannová and Pavlovský [50].*

Table 2.
Summary of adverse effects on soils due to the accumulation of metal(loid)s.†

[17]. Moreover, natural attenuation is often ineffective to eliminate the excess metal(loid)s from the soil, while the remediation process requires high cost and long duration in most instances [13]. Hence, it is necessary to estimate the variation in metal(loid) abundances of soils, which are susceptible to anthropogenic exposure, continuously or even periodically to avoid foreseeable mandatory soil cleanup requirements. The protocols for the assessment of metal(loid) contamination of soils will be discussed in the current chapter, preceded with a brief overview of the sources and toxicity impacts of metal(loid)s in soils.

2. Potentially toxic metal(loid)s

Metal(loid)s, which are ubiquitous in natural soil, and described to have influence on the physiological functions of living beings, e.g., plants, and other organisms, can be classified as nutritionally essential, nonessential with a possible beneficial effect, or nonessential with no beneficial effects [18] as listed in **Table 3**. The nonessential elements are potentially toxic even at deficient concentrations, while the essential ones can exert harmful impacts at elevated levels [19]. Metal(loid)s, those evoke health concerns, when accumulated in soils, exert chronic toxic effects on humans and other living beings usually via food-chain transfer. However, acute metal(loid) poisoning, even though rare, might also occur through ingestion, inhalation, or dermal contact. The toxicokinetics and toxicodynamics of metal(loid)s depend on several factors, e.g., route of exposure, dose, chemical speciation, solubility, and biotransformation, including the age, gender, and nutritional status of the exposed individuals [20]. Moreover, co-exposure to metal(loid)s mixtures may produce additive, antagonistic, or synergistic toxic effects, which could be more severe at both relatively high-dose and low-dose levels [21, 22].

An analysis of published data indicates that As, Cd, Cr, Pb, and Hg are systemic toxicants among the metal(loid)s [20], which are known to induce adverse health effects in humans ranging from dermatological, gastrointestinal, neurologic, hematologic, immunologic, metabolic, nephrotic, developmental, and behavioral disorders to cancers [23–25]. The As, Cd, Cr, Pb, or Hg might also interfere

Nutritionally essential metal(loid)s	Metal(loid)s with possible beneficial effects	Metal(loid)s with no known beneficial effects
Cobalt	Boron	Aluminum
Chromium(III)	Nickel	Antimony
Copper	Silicon	Arsenic
Iron	Vanadium	Barium
Manganese		Beryllium
Molybdenum		Cadmium
Selenium		Lead
Zinc		Mercury
		Silver
		Strontium
		Thallium

†*Source: Goyer et al. [18].*

Table 3.
Classification of metal(loid)s based on the health impact characteristics.†

metabolically with the nutritionally essential metal(loid)s, such as Fe, Ca, Cu, and Zn [26, 27]. The ecotoxicological considerations expanded the list of hazardous elements including a total of 11 metal(loid)s (As, Ba, Cd, Cr, Cu, Hg, Ni, Sb, Se, Tl, and V) [28]. The US-EPA priority pollutant list [29], however, included Ag, Be, Pb, and Zn in the list of toxic metal(loid)s and excluded Ba and V.

3. Assessment of soil contamination by metal(loid)s

A soil system is "contaminated" if any or more than a few of the toxic metal(loid) are present where it should not be or above the designated "background" concentrations [30, 31]. However, the definition of the term "background" is yet to be defined universally [6], and a selective list of definitions used to define the "background" conditions are listed in **Table 4**. A critical evaluation of "background" definitions [32] revealed that a precise global background value for an individual metal(loid) could not be proposed because there have been ups and downs in the overall natural distribution metal(loid)s in the ecosphere. Hence, it should be limited to specific geographic locations or regions and should be considered as a range instead of an absolute value to deal with the unavoidable environmental heterogenicity [32–34]. The regional "background" values of metal(loid)s represent either off-site or on-site reference locations. The off-site "background" values, as derived from real sample measurements [35, 36], often do not have sufficient metadata to validate the data accuracy [37] and also do not include the impact of transboundary atmospheric transport of metal(loid)s [38, 39]. The on-site "background" values usually represent buried fossil topsoils [40], dated peat bog samples [41], or deep soil layer from the same soil profile [42, 43]. However, the buried topsoils might subsequently be depleted by pedogenetic processes [44], and the properties of deep soil layers, e.g., organic matter content, bulk density, and so forth, are different from those of top soils [39, 45]. *Clarke values* are used as the representative "background" when regional off-site or on-site reference data is not available or cannot be obtained [6, 36]. *Clarke values*, even though used as an arbitrary off-site reference, does not sufficiently represent variations in element distributions in a regional or local context because of the lithologic discontinuities or pedogenic processes [34, 46]. The critical point is to select the correct "baseline" value to avoid mistaken identification of soil contamination that would create negative economic and social impacts. The strategies to avoid data bias in environmental monitoring of soil contamination are discussed by Desaules [37]. The distribution of geochemical data and related issues are focused in the works of Reimann and Filzmoser [47] and Reimann and de Caritat [45].

The methods used for soil contamination assessment include both statistical and geochemical methods, which are critically evaluated by several researchers, e.g., Desaules [39], Morrow et al. [48], D'Amore et al. [49], Weissmannová and Pavlovský [50], Cai et al. [51], Mizutani et al. [52], and so forth.

Definition	Reference
The concentration of a metal(loid) reflecting natural processes uninfluenced by human activities	[32]
The normal abundance of a metal(loid) in barren earth material	[6, 53]
Geogeneous or pedogeneous average concentration of a metal(loid) in an examined soil	[6, 54]

Table 4.
A selective list of definitions used to define "background" metal(loid) concentration in soils.

4. Conclusion

Metals in soil induce long-term risks to the ecosystems. Dynamics of metals in ecosphere can be assessed precisely using the information on the interactions of metals with environmental compartments. Evaluation of total metal content in soil and comparison with the "background" concentrations are the basic idea to deduce the anthropogenic inputs. However, there are differences in opinion regarding the test methods, definitions of "background," or approaches in data interpretation for the assessment of soil contamination. Hence, it might require more time to unify the understanding of soil contamination with metals.

Author details

Ismail M.M. Rahman[1*] and Zinnat A. Begum[2,3]

1 Institute of Environmental Radioactivity, Fukushima University, Fukushima City, Fukushima, Japan

2 Venture Business Laboratory, Organization of Frontier Science and Innovation, Kanazawa University, Kanazawa, Japan

3 Department of Civil Engineering, Southern University Bangladesh, Bangladesh

*Address all correspondence to: immrahman@ipc.fukushima-u.ac.jp

IntechOpen

© 2019 The Author(s). Licensee IntechOpen. This chapter is distributed under the terms of the Creative Commons Attribution License (http://creativecommons.org/licenses/by/3.0), which permits unrestricted use, distribution, and reproduction in any medium, provided the original work is properly cited.

References

[1] Kabata-Pendias A. Trace Elements in Soils and Plants. 4th ed. Boca Raton, FL: CRC Press; 2010

[2] Alloway BJ. Sources of heavy metals and metalloids in soils. In: Alloway BJ, editor. Heavy Metals in Soils: Trace Metals and Metalloids in Soils and their Bioavailability. Dordrecht: Springer Netherlands; 2013. pp. 11-50

[3] Hans Wedepohl K. The composition of the continental crust. Geochimica et Cosmochimica Acta. 1995;**59**:1217-1232

[4] Tan L, Chi-lung Y. Abundance of chemical elements in the earth's crust and its major tectonic units. International Geology Review. 1970;**12**:778-786

[5] Taylor SR, McLennan SM. The geochemical evolution of the continental crust. Reviews of Geophysics. 1995;**33**:241-265

[6] Wu J, Teng Y, Lu S, Wang Y, Jiao X. Evaluation of soil contamination indices in a mining area of Jiangxi, China. PLoS One. 2014;**9**:e112917

[7] Sposito G, Page AL. Cycling of metal ions in the soil environment. In: Sigel H, Sigel A, editors. Metal Ions in Biological Systems. Circulation of Metals in the Environment. Vol. 18. New York: Marcel Dekker; 1984. pp. 287-332

[8] Adriano DC. Trace Elements in Terrestrial Environments: Biogeochemistry, Bioavailability, and Risks of Metals. New York: Springer; 2001

[9] Kirpichtchikova TA, Manceau A, Spadini L, Panfili F, Marcus MA, Jacquet T. Speciation and solubility of heavy metals in contaminated soil using X-ray microfluorescence, EXAFS spectroscopy, chemical extraction, and thermodynamic modeling. Geochimica et Cosmochimica Acta. 2006;**70**:2163-2190

[10] Young SD. Chemistry of heavy metals and metalloids in soils. In: Alloway BJ, editor. Heavy Metals in Soils: Trace Metals and Metalloids in Soils and their Bioavailability. Dordrecht: Springer Netherlands; 2013. pp. 51-95

[11] Chlopecka A, Bacon JR, Wilson MJ, Kay J. Forms of cadmium, lead, and zinc in contaminated soils from southwest Poland. Journal of Environmental Quality. 1996;**25**:69-79

[12] Su C, Jiang L, Zhang W. A review on heavy metal contamination in the soil worldwide: Situation, impact and remediation techniques. Environmental Skeptics and Critics. 2014;**3**:24-38

[13] Wood JM. Biological cycles for toxic elements in the environment. Science. 1974;**183**:1049-1052

[14] Yongsheng Q. Study on the influences of combined pollution of heavy metals Cu and Pb on soil respiration. Journal of Anhui Agricultural Sciences. 2008;**36**:1117-1128

[15] Dermont G, Bergeron M, Mercier G, Richer-Laflèche M. Metal-contaminated soils: Remediation practices and treatment technologies. Practice Periodical of Hazardous, Toxic, and Radioactive Waste Management. 2008;**12**:188-209

[16] Dermont G, Bergeron M, Mercier G, Richer-Lafleche M. Soil washing for metal removal: A review of physical/chemical technologies and field applications. Journal of Hazardous Materials. 2008;**152**:1-31

[17] Alekseenko V, Alekseenko A. The abundances of chemical elements in

urban soils. Journal of Geochemical Exploration. 2014;**147**:245-249

[18] Goyer R, Golub M, Choudhury H, Hughes M, Kenyon E, Stifelman M. Issue paper on the human health effects of metals. Risk Assessment Forum. Washington, DC: U.S. Environmental Protection Agency; 2004

[19] Alloway BJ. Heavy Metals in Soils: Trace Metals and Metalloids in Soils and their Bioavailability. Dordrecht: Springer Netherlands; 2013

[20] Tchounwou PB, Yedjou CG, Patlolla AK, Sutton DJ. Heavy metal toxicity and the Environment. In: Luch A, editor. Molecular, Clinical and Environmental Toxicology: Volume 3: Environmental Toxicology. Basel: Springer Basel; 2012. pp. 133-164

[21] Wang G, Fowler BA. Roles of biomarkers in evaluating interactions among mixtures of lead, cadmium and arsenic. Toxicology and Applied Pharmacology. 2008;**233**:92-99

[22] Nordberg GF, Jin T, Hong F, Zhang A, Buchet JP, Bernard A. Biomarkers of cadmium and arsenic interactions. Toxicology and Applied Pharmacology. 2005;**206**:191-197

[23] IARC. Report on the 'Meeting of the International Agency for Research on Cancer (IARC) working group on beryllium, cadmium, mercury and exposures in the glass manufacturing industry' Scandinavian Journal of Work, Environment and Health. 1993;**19**:360-363

[24] Begum ZA, Rahman IMM, Hasegawa H. Soil Remediation with Green Chelants. Scholars' Press. Germany: Saarbrücken; 2015

[25] NTP (National Toxicology Program). "Report on carcinogens". North Carolina: U.S. Department of Health and Human Services Research Triangle Park; 2016

[26] Alonso ML, Montaña FP, Miranda M, Castillo C, Hernández J, Benedito JL. Interactions between toxic (As, Cd, Hg and Pb) and nutritional essential (Ca, Co, Cr, Cu, Fe, Mn, Mo, Ni, Se, Zn) elements in the tissues of cattle from NW Spain. Biometals. 2004;**17**:389-397

[27] Abdulla M, Chmielnicka J. New aspects on the distribution and metabolism of essential trace elements after dietary exposure to toxic metals. Biological Trace Element Research. 1989;**23**:25-53

[28] Vodyanitskii YN. Contamination of soils with heavy metals and metalloids and its ecological hazard (analytic review). Eurasian Soil Science. 2013;**46**:793-801

[29] US EPA. The Clean Water Act (Priority Pollutants: Appendix A to 40 CFR Part 423). Washington, D.C.: United States Environmental Protection Agency; 2012

[30] US EPA. Terms of Environment: Glossary, Abbreviations, and Acronyms. Washington, D.C.: United States Environmental Protection Agency; 1997

[31] Chapman PM. Adaptive monitoring based on ecosystem services. Science of the Total Environment. 2012;**415**:56-60

[32] Reimann C, Filzmoser P, Garrett RG. Background and threshold: Critical comparison of methods of determination. Science of the Total Environment. 2005;**346**:1-16

[33] Edelman T, de Bruin M. Background values of 32 elements in Dutch topsoils, determined with non-destructive neutron activation analysis. In: Assink JW, Van Den Brink WJ, editors. Contaminated Soil. Dordrecht: Springer Netherlands; 1986. pp. 89-99

[34] Anderson RH, Kravitz MJ. Evaluation of geochemical

associations as a screening tool for identifying anthropogenic trace metal contamination. Environmental Monitoring and Assessment. 2010;**167**:631-641

[35] De Temmerman LO, Hoenig M, Scokart PO. Determination of "normal" levels and upper limit values of trace elements in soils. Zeitschrift für Pflanzenernährung und Bodenkunde. 1984;**147**:687-694

[36] Ikem A, Campbell M, Nyirakabibi I, Garth J. Baseline concentrations of trace elements in residential soils from Southeastern Missouri. Environmental Monitoring and Assessment. 2008;**140**:69-81

[37] Desaules A. The role of metadata and strategies to detect and control temporal data bias in environmental monitoring of soil contamination. Environmental Monitoring and Assessment. 2012;**184**:7023-7039

[38] Bindler R, Brännvall M-L, Renberg I, Emteryd O, Grip H. Natural lead concentrations in pristine boreal forest soils and past pollution trends: A reference for critical load models. Environmental Science & Technology. 1999;**33**:3362-3367

[39] Desaules A. Critical evaluation of soil contamination assessment methods for trace metals. Science of the Total Environment. 2012;**426**:120-131

[40] Elberling B, Breuning-madsen H, Hinge H, Asmund G. Heavy metals in 3300-year-old agricultural soils used to assess present soil contamination. European Journal of Soil Science. 2010;**61**:74-83

[41] Shotyk W, Blaser P, Grünig A, Cheburkin AK. A new approach for quantifying cumulative, anthropogenic, atmospheric lead deposition using peat cores from bogs: Pb in eight Swiss peat bog profiles. Science of the Total Environment. 2000;**249**:281-295

[42] Blaser P, Zimmermann S, Luster J, Shotyk W. Critical examination of trace element enrichments and depletions in soils: As, Cr, Cu, Ni, Pb, and Zn in Swiss forest soils. Science of the Total Environment. 2000;**249**:257-280

[43] Bourennane H, Douay F, Sterckeman T, Villanneau E, Ciesielski H, King D, et al. Mapping of anthropogenic trace elements inputs in agricultural topsoil from Northern France using enrichment factors. Geoderma. 2010;**157**:165-174

[44] ISO. ISO 19258: 2005 (Soil quality—Guidance on the determination of background values). Geneva: International Organization for Standardization; 2005

[45] Reimann C, de Caritat P. Distinguishing between natural and anthropogenic sources for elements in the environment: Regional geochemical surveys versus enrichment factors. Science of the Total Environment. 2005;**337**:91-107

[46] Salminen R, Gregorauskien V. Considerations regarding the definition of a geochemical baseline of elements in the surficial materials in areas differing in basic geology. Applied Geochemistry. 2000;**15**:647-653

[47] Reimann C, Filzmoser P. Normal and lognormal data distribution in geochemistry: Death of a myth. Consequences for the statistical treatment of geochemical and environmental data. Environmental Geology. 2000;**39**:1001-1014

[48] Morrow DA, Gintautas PA, Weiss AD, Piwoni MD, Bricks RM. Metals Speciation in Soils: A Review of Methodologies (Technical Report IRRP-96-5). Washington, DC: U.S. Army Corps of Engineers; 1996

[49] D'Amore JJ, Al-Abed SR, Scheckel KG, Ryan JA. Methods

for speciation of metals in soils. Journal of Environmental Quality. 2005;**34**:1707-1745

[50] Weissmannová HD, Pavlovský J. Indices of soil contamination by heavy metals—Methodology of calculation for pollution assessment (minireview). Environmental Monitoring and Assessment. 2017;**189**:616

[51] Cai C, Xiong B, Zhang Y, Li X, Nunes LM. Critical comparison of soil pollution indices for assessing contamination with toxic metals. Water, Air, and Soil Pollution. 2015;**226**:352

[52] Mizutani S, Ikegami M, Sakanakura H, Kanjo Y. Test methods for the evaluation of heavy metals in contaminated soil. In: Hasegawa H, Rahman IMM, Rahman MA, editors. Environmental Remediation Technologies for Metal-Contaminated Soils. Tokyo: Springer Japan; 2016. pp. 67-97

[53] Hawkes HE, Webb JS. Geochemistry in Mineral Exploration. New York: Haper & Row; 1962

[54] ISO. ISO 11074: 2015 (Soil quality—Vocabulary). Geneva: International Organization for Standardization; 2015

Chapter 2

The Influence of Potentially Toxic Elements on Soil Biological and Chemical Properties

Om Prakash Bansal

Abstract

Soil has been a source of wealth for humans for infinite years and it continues so at present. Both mineral and organic amendments have been applied to soil to slow down its progressive impoverishment. Biological activity, mainly microbial activity, plays a key role in the stability and fertility as well as in biogeochemical cycles. Effect of potentially toxic elements on soil microbial activity, the composition of soil microbial community, soil enzyme activities, and soil physiochemical properties have been reviewed in this work.

Keywords: potentially toxic elements, plant growth, soil microbial activity, soil microbial composition, soil enzyme activities, soil physicochemical properties

1. Introduction

The soil is the basic source for the human being living and most fundamental elements of human production and the carrier linking human economic relationship together. Contamination of soils by potentially toxic elements due to different anthropogenic activities is one of the factors which influence the life in soils [1–3]. There are four major pathways [4] by which potentially toxic elements enter in the soils: (i) atmosphere to soils, (ii) sewage to soils, (iii) solid waste to soil, and (iv) agricultural supplies to soils. Potentially toxic elements are the elements with high density and high relative atomic weight, showing metallic properties as ductility, malleability, conductivity, ligand specificity [5]. Some potentially toxic elements such as Co, Cu, Fe, Mn, Mo, Ni, and Zn are beneficial to the biological system when present in permissible amount but damage the biological system if present in excess. Soil potentially toxic elements such as Pb, Cd, Hg and As (a metalloid but generally referred to as a potentially toxic element) are harmful to crops, humans, and animals. Potentially toxic elements are added to the soil naturally and by anthropogenic activities; metals Cd, Pb, Zn and Ni are also originated from heavy traffic on roads and causes soil pollution. Oves et al. [6] reported that the annual estimate of potentially toxic elements release from all sources in worldwide is around 22,000 metric ton of Cd; 939,000 of Cu; 783,000 of Pb and 1,350,000 of Zn. As the soil and potentially toxic elements have rich and diverse binding characteristics, soil acts as a major reservoir media for potentially toxic elements. Depending on several factors such as water content, pH, temperature, particle size, clay content and the nature of potentially toxic element, the soil matrix may adsorb, exchange, oxidize, reduce, and catalyze the metal ions. The mobility and toxicity of potentially toxic elements in soils depend on the soil-metal composition and metal ion concentration. The availability of potentially toxic elements in soils besides other factors also depends on soil density and type of charge in soil colloids, soil surface area and the power of complexion with ligands [7, 8].

The present study was undertaken to understand the current situation and the impact of potentially toxic elements on human, on soil microbial activity, on soil microbial composition on soil enzyme activities and on soil physicochemical properties.

2. Impact of potentially toxic elements on plant growth

In soils, plants can uptake potentially toxic elements which are water-soluble or get easily solubilized by roots [9]. As potentially toxic elements cannot be degraded, when their concentrations within the plant exceed permissible limit they adversely affect the plant directly and indirectly. Leaf chlorosis, disturbed water balance, and reduced stomatal opening, inhibition of cytoplasmic enzymes and damage to cell structures due to oxidative stress [10], are some direct toxic effects of potentially toxic elements on plants. Replacement of essential nutrients at cation exchange sites of plants by potentially toxic elements is one of the examples of indirect toxic effect [11]. High metal concentration may lead to decrease in organic matter decomposition, the decline in soil nutrients, decrease in enzymatic activities useful for plant metabolism lead to a decline in plant growth which sometimes results in the death of plant [12].

3. Impact of potentially toxic elements on human

The uptake of potentially toxic elements from contaminated soils by plants comprises a major path for these elements to enter the human and animal food chain [13]. Potentially toxic elements' bioaccumulated in the food chain, harmful to human, enters the human body through inhalation and ingestion and ingestion is the main route of accumulation in humans. Besides it, the potentially toxic elements have also been used for a long time by humans for making metal alloys and pigments for paints, cement, paper, rubber, and other materials. These potentially toxic elements enter the body system when these plants are directly or indirectly consumed, also through air and water and may bioaccumulate over a period of time [14, 15]. Potentially toxic elements entered in the human body by any means affect the immune system, basic physiological processes of cell and gene expression and may cause nausea, anorexia, vomiting, gastrointestinal abnormalities, and dermatitis [16, 17]. Women are more susceptible to the adverse effects of Cd and have higher body burdens due to the long half-life of Cd and increased dietary absorption of Cd is long-lived in the body and low-level cumulative exposure has been associated with changes in renal function and bone metabolism [18]. Potentially toxic elements mainly lead (Pb) effects and damages body organs and systems as kidney [19]; liver [20]; and change blood composition, [21]; damage lungs [22]; reproductive system [23]; central nervous system [4]; urinary system [24]; immune system. Chen [25] has reported that workers who are in close contact with nickel powder are more likely to suffer from respiratory cancer and nasopharyngeal carcinoma. At lower concentration copper acts as co-factors for various enzymes of redox cycling [26]; however, the higher concentration of Cu disrupts the human metabolism leading to anemia, liver and kidney damage, stomach and intestinal irritation. Arsenic induces skin, liver, lung, colotral uterine cancers [27].

Physical, muscular, and neurological impairments, the degenerative processes similar to Alzheimer's disease [19], Parkinson's disease [28], muscular dystrophy and multiple sclerosis [29] may occur when the human poulation is exposed to potentially toxic elements for a long time.

4. Effects of potentially toxic elements on soil microbial activity

Potentially toxic elements are common and important refractory pollutants, affect the number, diversity and microbial activity of soil microorganisms. As soil microorganisms decompose organic matter, affects nutrient cycling, microbes play a major role in soil fertility and primary production. Soil microbial biomass which is useful for studying the harmful effects of toxic metals at the cellular level is mainly made up of bacteria and fungi. The toxic effect of potentially toxic elements depends on the number of metals bioaccumulated by absorption, migration, and transformation. Adverse effects of higher concentration of toxic metals on microorganisms may be due to inactivation of enzyme activity center; electron-donating groups such as mercapto protein, nucleic acid base and phosphate combination, accumulation of toxic metals more than the ability of organisms to bear resulting in biological disease and death; inhibition in the formation of metallothionein or metalloprotein. A number of studies have shown that the higher number of metallothionein in cells induces the anti-apoptotic effects, and a decrease in the number of metallothionein increases the susceptibility to apoptotic cell death.

Soil microbes are the main participant of all the soil biochemical processes. Soil biochemical processes are the tools for maintaining soil quality; formation of soil organic matter; decomposition of harmful substances; formation of soil structure and biochemical cycles. Contamination of soils by toxic metals decreases soil microbial properties such as soil respiration, enzymatic activities. Soil microbial properties depend on soil pH, organic matter and other chemical properties. Severe potentially toxic element pollution can inhibit soil microbial activity and seriously threaten the soil ecosystem function.

A number of workers [30–33] have reported that potentially toxic elements particularly cadmium, copper, and zinc can disrupt the microbiological equilibrium of soil. Disturbances of the biological balance of soil caused by the excess of potentially toxic elements might be attributed to the disruption of physiological functions, denaturation of proteins and destruction of cellular membranes of soil microorganisms [34–36]. Potentially toxic elements immobilize soil bacteria, while microbial metabolites enhanced the mobility of potentially toxic elements [37–39]. Potentially toxic elements in different quantities and forms in soils cause changes in the counts of microorganisms, microbial biomass and microbial activity [33, 40–43] via inhibiting microbial community diversity, particularly that of fungal groups i.e., Ascomycota and Chytridiomycota) in the large-size fractions, which mainly depends on heterogeneous SOC availability across the PSFs. Potentially toxic elements create abiotic stresses [37, 38] by inducing disorders in the metabolism of the microorganism. Damages to the control systems regulated by regulatory and signal proteins, including the cell's development, apoptosis and regulation of the cellular cycle are caused by potentially toxic elements [36], which might be due to the blocking of enzymatic active centers and driving away cations that are important for the functioning of a cell, supplanting their functions, e.g., discontinuation of the cell-to-cell adhesion (cadmium), direct binding with the DNA (chromium), interacting with the binding sites of protein phosphatases (vanadium) [44]. Tolerant species of microbes demonstrate higher resistance to stress factors than sensitive ones [45]. Tolerance of potentially toxic elements is associated with [46–48]: (1) specific transport of metal ions in the cytoplasmic membrane; (2) synthesis and excretion to the environment chelating compounds, which bind and transport ions dissolved in the environment; (3) sorption of ions onto mucosal surfaces and the binding by biopolymers of the wall and membrane complex; (4) the presence of plasmids in a bacterial cell, which enables it to acquire resistance to toxic elements.

Many researchers [49–51] have reported that when potentially toxic elements are present in the excessive amount in the soil the microbial count and diversity of microorganisms' decreases. Bansal and Mishra [52] reported that there was a significant increase in the bacterial and fungal population and decrease in actinomycetes population in sewage irrigated soils. The population density of bacteria and fungi increased with duration of sewage irrigation up to 14 days of incubation and thereafter decreased. Kouchou [53] during their studies found that potentially toxic elements contamination in alkaline soils has a negative effect on actinomycetes and fungi soil populations, while a positive effect on the total aerobic heterotrophic bacterial population. They also inferred that the effect of potentially toxic elements on microbial population of the soil is dependent on several factors related to soil environment and soil physicochemical characteristics. Fengqiu et al. [54] during their studies on Microbial diversity and community structure in agricultural soils suffering from 4 years of Pb contamination found that the presence of Pb^{2+} in soil showed the weak impact on the diversity of soil bacterial community. Contamination of soil by Pb influences soil chemical properties and number of some genera of bacteria. The number of heavy metal-resistant bacteria at genus level viz. *Bacillus*, *Streptococcus*, and *Arthrobacter* is highly correlated with the amount of Pb. There was a negative correlation between soil organic matter and available Pb and total Pb. The total relative abundance of Gemmatimonadetes, Nitrospirae, and Planctomycetes was negatively correlated with total Pb. Both the microbial community composition and physicochemical properties of soil are influenced by the amount of Pb. Workers [52, 55] found that nitrifying bacteria, symbiotic nitrogen-fixing bacteria, and *Azotobacter* spp. are the microorganisms most susceptible to potentially toxic elements. Potentially toxic elements produce a stronger effect on *Azotobacter* cells than organotrophic bacteria mainly because richer communities of microbes are more resistant to potentially toxic elements than single species and genera [51, 56]. Wyszkowska et al. [48] reported that the inhibitory effect of potentially toxic elements on soil microorganisms can be represented as follows: oligotrophic bacteria: (Ni > Pb > Cr(III) > Cu > Zn > Cd), copiotrophic bacteria: (Cd > Ni > Cr(III) > Zn > Cu), ammonifying bacteria: (Ni > Pb > Cr(III) > Cd > Zn > Hg), nitrogen immobilizing bacteria: (Zn > Cr(III) > Hg > Cu), actinomycetes: (Cu > Cr(III) > Ni > Zn > Pb). Few researchers [57] found that crops can moderate the influence of potentially toxic elements on soil microbes, crops improve the microbiological activity of the soil, mainly owing to substances secreted by roots.

5. Effects of potentially toxic elements on soil microbial composition

Potentially toxic elements when accumulated in soil beyond their permissible limit they firstly influence the quality and quantity of soil bacteria, fungi, actinomycetes, and other microbial population. Potentially toxic element contamination in soil not only produces different microbial community patterns but also change the chemical and biological properties of the soil. In the soils which are polluted by potentially toxic elements for a long time, those soil microorganisms which can specifically be adapted exist. The efficiency of microbial populations in organic mineralization is inversely correlated with the soil organic carbon content, an indicator of the impact of potentially toxic element pollution. Microbial communities in soils are very important as they are helpful in nutrient cycling, plant symbiosis, and the detoxification of noxious chemicals (used to control plant pests and plant growth) [58]. When metal enriched sewage sludge is added to soils microbial biomass leads to a decrease in functional diversity [59] and changes in microbial community structure [60]. However, metal exposure may also lead to the development of metal-tolerant microbial population [61].

Potentially toxic elements affect the microbial activity and microbial community structure. The number of bacteria and actinomycetes significantly decreased as compared to the control, while no significant difference in fungal cells up to 2 weeks of incubation [62] of potentially toxic elements. The negative effect of studied potentially toxic elements on culturable heterotrophic bacteria was more than actinomycetes and fungi. The DGGE profile indicated that the bacterial community structure was changed in the Cd/Pb-amended samples, particularly at high concentrations, but not bacterial taxon richness and community composition [63]. The relative abundance of specific bacterial taxa including, Acidobacteria, Actinobacteria, Bacteroidetes, Chloroflexi, Planctomycetes, and Probacteria is affected by potentially toxic elements pollution. A significant correlation between a group of metal-resistance genes and metal concentration in soil was also reported by Azarbad et al. [63]. Acidobacteria Gp and *Proteobacteria thiobacillus* bacteria had more links between nodes and more positive interactions among microbes in CL- and CH-networks were positively correlated with cadmium concentration while *Longilinea*, Gp2 and Gp4, had fewer network links and more negative interactions in CL and CH-networks where negatively correlated with Cd [63]. There was the only positive correlation in between the members of the phyla Crenarchaeota and Euryarchaeota, class Thermoprotei and order Thermoplasmatales and Cd metal and had more network interactions in CH-networks. Li et al. [64] also reported that (i) the microbial community composition, as well as a network interaction was the shift to strengthen adaptability of microorganisms to potentially toxic element contamination, (ii) archaea was resistant to potentially toxic element contamination and may contribute to the adaption to potentially toxic elements. Pb^{2+} in soil showed a weak impact on the diversity of soil bacteria community, but it influenced the abundance of some genera of bacteria, as well as soil physicochemical properties [54]. At the genus level, there was a significant difference in the relative abundances of heavy-metal-resistant bacteria such as *Bacillus*, *Streptococcus*, and *Arthrobacter*. The abundance of main bacteria phyla was highly correlated with total Pb. The relative abundance of Gemmatimonadetes, Nitrospirae, and Planctomycetes was negatively correlated with total Pb. Lead influences both the microbial community composition and physicochemical properties of soil.

6. Effects of potentially toxic elements on soil enzyme activities

The biological activity of soils is an essential parameter of their ecological status. The potentially toxic elements present in the soil due to anthropogenic activities disturb the normal functioning of soil biota and, hence, the entire soil system. As the concentration of potentially toxic elements increases, the activity of most enzymes is significantly reduced and may be caused directly by the interaction between the enzyme and the potentially toxic elements, which is not associated with a reduction in microbes. Potentially toxic elements influence the enzymatic activity, by destroying the spatial structure of the active groups of the enzyme, by inhibiting the growth and reproduction of microorganisms which reduces the synthesis and metabolism of the microbial enzymes. Soil microbes and soil enzymatic activities are significantly correlated. Some enzymes secreted by microorganisms participate in soil ecosystems and energy together.

Potentially toxic elements in enzymes play a triple function: catalytic, structural and regulatory. Zinc is an integral part of the number of enzymes and number of intracellular enzymes viz., carbon anhydrase, carboxypeptidase, thermolysin, alkaline phosphatase, dehydrogenases (glyceraldehyde-3-phosphate, alcohol, glutamine), fructo-diphosphate aldolases, superoxide dismutase, DNA and RNA polymerase, tRNA transferase need zinc for proper functioning. When potentially toxic elements are present in excess the natural functions of metals are distorted.

Tejada et al. [65]; Yu and Cheng, [66] reported that the enzyme reactions are inhibited by potentially toxic elements in three different ways: (1) complexation of the substrate; (2) combination with protein-active groups on the enzyme, and; (3) reaction with the enzyme-substrate complex, while Vig et al. [67] reported that as potentially toxic elements interact with the enzyme active sites and substrate complexes, denatures the enzyme protein and competes with metal ions those are needed to form enzyme-substrate complexes, inhibit soil enzyme activities. Nuaimi and Maktoom [68] reported that potentially toxic element pollutants found in the soil can cause their deleterious effects by any of four ways: (1) There is oxidative stress in organisms due to the release of the oxy radical which is produced by redox cycling of potentially toxic elements, (2) proteins are deactivated or denatured as metals bind to sulfhydryl groups of proteins, (3) antioxidant ability of cells retards as potentially toxic elements bind an intracellular glutathione and/or antioxidant enzymes, (4) metalloenzymes became inactivated as potentially toxic elements compete for metal cofactor binding of metalloenzymes. Nuaimi and Maktoom [68] also reported that the potentially toxic elements such as Hg^{2+}, Cu^{2+} inhibited alkaline phosphatase enzyme more strongly than Cd^{2+}, and Co^{2+} and also that alkaline phosphatase is readily inactivated with the exposure to potentially toxic elements, uncharged at neutral pH, arsenic can diffuse across the cell membrane, it can inhibit the enzymatic activities even at a low concentrations.

Bhattacharyya et al. [69] reported that arsenic reacts with the sulfhydryl group of the enzyme to form arsenic sulfide resulting in the decrease of enzymatic activities. The decrease in enzymatic activity by arsenic may be due to: (1) by interacting with the enzyme-substrate complex; (2) by denaturing the enzyme protein, or; (3) interacting with the active protein groups.

The influence of potentially toxic elements on soil enzyme activity depends on pH, nutrient form and amount, potentially toxic element concentration and availability, enzyme type, etc.

Ofoegbu et al. [70] during their research found that there was the significant negative correlation with the potentially toxic element contents and the activities of dehydrogenase, polyphenol oxidase, hydrogen peroxidase, alkaline and acid phosphatases and urease. There was a negative but non-significant correlation between Zn content and dehydrogenase activity; Cd content and hydrogen peroxidase and urease activities.

Diana et al. [71] reported that at the brownfield LSP, extracellular soil enzyme activities are notably high at a site with the highest potentially toxic element loads (soil had been unmanaged for over 40 years, left alone to naturally succeed), and there is a strong relationship between these enzyme activities and Cr and V in particular. This study demonstrates the capacity of some potentially toxic element contaminated soils to enzymatically function well under seemingly restrictive conditions.

Wiatrowska et al. [72] found that Dehydrogenases, acid and alkaline phosphatases exhibited the highest sensitivity toward Zn and it decreased in the order of metal concentrations: Zn > Cd > Cu > Pb. In contrast, urease was more tolerant to Zn. The sensitivity of urease was as follows: Cu > Zn > Cd > Pb. In respect of their sensitivity to concentrations of the bioavailable pool of Cd, Cu,

Pb, and Zn, the enzymes can be arranged as follows: alkaline phosphatase > acid phosphatase > dehydrogenases > urease. Bansal et al. [42] and Bansal [43] during their studies found that potentially toxic elements accumulated in soils due to sewage water irrigation increased the activity of the enzymes dehydrogenase, acid and alkaline phosphatase, urease and catalase up to 14 days of incubation and decreased thereafter.

Derdzyan et al. [73] during their studies found that potentially toxic element pollution affects the activities of beta-glucosidase, chitinase, leucine-aminopeptidase acid, phosphomonoesterase and acetate-esterase enzymes in the soils.

Gromakova et al. [74] during their work reported that the increase of mobile potentially toxic element forms content in soil inhibited the cellulose-degrading and urease activities after 30 days of the input of metal into the soil. The inhibition of biological activity of the studied metals formed the following series: Cd > Pb > Zn > Cu.

Yu and Cheng [66] during their studies found that addition of Cu, Cd, and Pb firstly enhanced urease activity and thereafter it declines, while with the increased concentration of Zn the activity of urease declines. In addition to Cu the catalase enzymatic activity initially enhanced and thereafter decreases, the catalase activity continuously decreases with the addition of Cd, Pb, and Zn. Addition of Cu, Cd, Pb, and Zn in soil results in a decrease in microbial biological carbon content.

Khan et al. [62] found that in the Cd and Pb treated soils the activity of acid phosphatase and urease was minimum after 2 weeks of incubation.

The enzymatic activities of the acid phosphatase and β-glycosidase were lowest in the soil samples having the maximum concentration of potentially toxic elements [75]. They also reported activity of acid phosphatase and β-glycosidase was significantly influenced by pH and non-significantly with soil organic matter.

A review of the literature shows that sensitivity of: dehydrogenases [76, 77] is: Hg (2 mg) > Cu (35 mg) > Cr^{6+} (71 mg) > Cr^{3+} (75 mg) > Cd^{2+} (90 mg) > Ni^{2+} (100 mg) > Zn^{2+} (115 mg) > $As3^{+}$ (168 mg) > Co^{2+} (582 mg) > Pb^{2+} (652 mg kg^{-1}), Cu^{2+} > Zn^{2+} > $Cr6^{+}$ > Hg^{2+} > Ni^{2+} > Cd^{2+} > Cr^{3+},: acid phosphatase: Cu^{2+} > Al^{3+} > Cd^{2+} > Zn^{2+} > Fe^{3+} > Ni^{2+} > Pb^{2+} > Sn^{2+} > Fe^{2+} > Co^{2+}, and alkaline phosphatase: Cd^{2+} > Al^{3+} > Zn^{2+} > Fe^{3+} > Cu^{2+} > Pb^{2+} > Ni^{2+} > Fe^{2+} > Se^{2+} > Co^{2+}.

7. Effects of potentially toxic elements on soil physicochemical properties

Li et al. [78] during their studies found that water holding capacity, soil bulk density, porosity, permeability, infiltration besides other factors also depends on the concentration of potentially toxic elements. Soil chemical properties depend on soil pH which affects the availability of soil nutrients and form of potentially toxic elements. The amount of plant available organic matter is also influenced by the concentration of potentially toxic elements.

Dawaki et al. [79] during their studies on the effects of heavy metals on physicochemical properties found that clay content and soil pH were non-significantly negatively correlated with soil total potentially toxic element's concentration. Organic carbon, cation exchange capacity, total nitrogen, phosphorous, calcium, potassium, sodium were positively significantly correlated with soil chromium, zinc and lead content, while no significant correlation with copper and nickel content. Sharma and Raju [80] reported that soil pH is positively correlated with potentially toxic elements content. Soil moisture content is positively correlated with potentially toxic elements content except for Cu and Cr. They also reported that there was no specific correlation between potentially toxic elements content and soil water holding capacity. There was a negative correlation among potentially toxic elements. Tripathi and Mishra [81] during their studies found that soil moisture content is positively significantly correlated with soil water holding capacity, soil organic matter, soil cation exchange capacity, amount of iron and lead; chromium content was significantly correlated with lead and nickel content while there was a significant positive correlation between soil copper content and nickel and zinc content. Lead and zinc in soils are significantly positively correlated. Similar results were reported by Singare et al. [82] during their studies on physicochemical properties and heavy metal content of the soil samples from Thane Creek of Maharashtra, India. Nwaogu [83] reported that in presence of Hg, Pb and Cd (100 $\mu g/dm^3$) soil physicochemical properties viz., moisture, phosphate, sulfate, chloride, calcium carbonate, total nitrogen and organic carbon were significantly changed.

8. Conclusion

Due to the persistent, toxic and non-biodegradable nature of potentially toxic elements, the problem of potentially toxic elements in the soil is a turning point for soil scientists. Potentially toxic elements contaminated soils, in general, are nutrient deficient and are likely to become barren in future. Soil microbes are very sensitive to potentially toxic elements. Microbial activities reflect changes in soil environment and are considered to be sensitive indicators in the soil. Microorganisms and their enzymes exert a beneficial effect in soil quality, plant growth as the activities of microbes increase nutrient availability and stimulate the degradation of pollutants. The potentially toxic element in soils also affects the soil physicochemical properties and depends on the form and amount of potentially toxic elements.

Author details

Om Prakash Bansal
Chemistry Department, D.S. College, Aligarh, India

*Address all correspondence to: drop1955@gmail.com

IntechOpen

© 2018 The Author(s). Licensee IntechOpen. This chapter is distributed under the terms of the Creative Commons Attribution License (http://creativecommons.org/licenses/by/3.0), which permits unrestricted use, distribution, and reproduction in any medium, provided the original work is properly cited.

References

[1] Prajapati SK, Meravi N. Potentially toxic element speciation of soil and *Calotropis procera* from thermal power plant area. Proceedings of the International Academy of Ecology and Environmental Sciences. 2014;**4**:68-71

[2] Zojiali F, Hassani AH, sayedi MH. Bioaccumulation of chromium by *Zea mays* in a waste water-irrigated soil. An experimental study. Proceeding of the International Academy of Ecology and Environmental Sciences. 2014;**4**:62-67

[3] Baishya K, Samra HP. Effect of agrochemicals application on accumulation of heavy metals on soil of different land uses with respect to nutrient status. Journal of Environmental Science Toxicology and Food Technology. 2014;**8**:46-54

[4] Su C, Jiang L, Zhang W. A review on heavy metal contamination in the soil worldwide: Situation, impact and remediation techniques. Environmental Skeptics and Critics. 2014;**3**:24-38

[5] Algreen M, Rein A, Legind CN, Amundsen CE, Karlson UG, et al. Test of tree core sampling for screening of toxic elements in soils from a Norwegian site. International Journal of Phytoremediation. 2012;**14**:305-319

[6] Oves M, Sagir Khan M, Huda Qari A, Nadeen Felemban M, Alkeelbi T. Heavy metals: Biological importance and detoxification strategies. Journal of Bioremediation & Biodegradation. 2016;**7**:334-348

[7] Marques APGC, Rangel AOSS, Castro PML. Remediation of heavy metal contaminated soil: Phytoremediation as a potentially promoting clean up technology. Critical Reviews in Environmental Sciences and Technology. 2009;**39**:622-654

[8] Raja Rajeswari T, Sailaja N. Impact of heavy metals on environmental pollution. Journal of Chemical and Pharmaceutical Sciences. 2014;**3**:175-181

[9] Blaylock MJ, Hauang JW. Phytoextraction of metals. In: Raskin L, Enslay BD, editors. Phytoremediation of Toxic Metals Using Plants to Clean Environment. New York: Wiley Interscience; 2000. pp. 53-70

[10] Jadia CD, Fulekar MH. Phytoremediation of heavy metals: Recent techniques. African Journal of Biotechnology. 2009;**8**:921-928

[11] Taiz L, Zeiger E. Plant Physiology. Sunderland, MA, USA: Sinauer Associates; 2002

[12] Schaller A, Diez T. Plant specific aspects of heavy metals uptake and comparison with quality standards for food and forage crops. In: Sauerbeck D, Lübben S, editors. Der Einfluß von festen Abfällen auf Böden, Pflanzen. Germany: KFA, Jülich; 1991. pp. 92-125. (German)

[13] Marshall FM, Holden J, Ghose C, Chisala B, Kapungwe E, Volk J, et al. Contaminated irrigation water and food safety for the urban and periurban poor: Appropriate measures for monitoring and control from field research in India and Zambia. In: Inception Report DFID Enkar R8160. UK: SPRU, University of Sussex. www.pollutionfood.net; 2007

[14] Mapanda F, Mangwayana EN, Nyamangara JK, Giller E. The effect of long-term irrigation using wastewater on heavy metals contents of soils under vegetables in Harare, Zimbabwe. Agriculture, Ecosystem and Environment. 2005;**107**:151-158

[15] Trueby P. Impact of heavy metals on forest trees from mining areas. In: International Conference on Mining and

the Environment; Sudbury, Ontario, Canada. 2003

[16] Chui S, Wong YH, Chio HI, Fong MY, Chiu YM, et al. Study of heavy metals poisoning in frequent users of Chinese medicines in Hong Kong and Macau. Phytotherapy Research. 2013;**27**:859-863

[17] Tchounwou PB, Yedjou CG, Patlolla AK, Sutton DJ. heavy metal toxicity and the environment. EXS. 2012;**101**:133-164

[18] Salt DE, Prince RC, Pickeriong IJ, Raskin I. Mechanisms of cadmium mobility and accumulation in Indian mustard. Plant Physiology. 1995;**109**:1427-1433

[19] Suzuki K, Yabuki T, Ono Y. Roadside *Rhododendron pulchrum* leaves as bio indicator of heavy metal pollution in traffic areas of Okayama, Japan. Environmental Monitoring and Assessment. 2009;**149**:133-141

[20] Sadik NA. Effects of diallyl sulfide and zinc on testicular steroidogenesis in cadmium-treated male rats. Journal of Biochemical and Molecular Toxicology. 2008;**22**:345-353

[21] Cope CM, Mackenzie AM, Wilde D, Sinclair LA. Effects of level and form of dietary zinc on dairy cow performance and health. Journal of Dairy Science. 2009;**92**:2128-2135

[22] Kampa M, Castanas E. Human health effects of air pollution. Environmental Pollution. 2008;**151**:362-367

[23] Lindemann MD, Cromwell GL, Monegue HJ, Purser KW. Effect of chromium source on tissue concentration of chromium in pigs. Journal of Animal Science. 2008;**86**:2971-2978

[24] WHO Europe. Levels of Lead in Children's Blood Fact Sheet 4.5. Copenhagen, Denmark: World Health Organization Regional Office for Europe; 2009

[25] Chen YF. Review of the research on potentially toxic element contamination of China's city soil and its treatment methods. In: China Population, Resources and Environment. 2011;**21**:536-539

[26] Farhan M, Khan HY, Oves M, Al-Harrasi A, Rehmani N, et al. Cancer therapy by catechins involves redox cycling of copper ions and generation of reactive oxygen species. Toxin. 2016;**8**:37

[27] Science for Environment Policy in Depth Report Soil Contamination: Impacts on Human Health. Report Produced for the European Commission DG Environment; Sept. 2013

[28] Guilarte TR. Manganese and Parkinson's disease: A critical review and new findings. Ciência & Saúde Coletiva. 2011;**16**:4549-4566

[29] WHO. Exposure to Arsenic: A Major Public Health Concern. Geneva: World Health Organization; 2010

[30] Olaniran AO, Balgobind A, Pillay B. Review bioavailability of heavy metal in soil: Impact on microbial biodegradation of organic compounds and possible improvement strategies. International Journal of Molecular Sciences. 2013;**14**:10197-10228

[31] Liu A, Wang H, Gao P, Xu H. Chemical fractionation of Cu and Zn and their impacts on microbial properties in slightly contaminated soils. International Journal of Agricultural Research, Innovation and Technology. 2013;**3**:20-25

[32] Markowicz A, Plaza G, Piotrowska-seget Z. Activity and functional diversity of microbial communities in long-term hydrocarbon and potentially

toxic element contaminated soils. Archives of Environmental Protection. 2016;**42**:3-11

[33] Shi W, Ma X. Effects of heavy metal Cd pollution on microbial activities in soil. Annals of Agricultural and Environmental Medicine. 2017;**24**:722-725

[34] Bastida F, Kandeler E, Moreno JL, Ros M, Garcia C, Hernandez T. Application of fresh and composted organic wastes modifies structure, size and activity of soil microbial community under semiarid climate. Applied Soil Ecology. 2008;**40**:318-329

[35] Friedlova M. The influence of heavy metals on soil biological and chemical properties. Soil and Water Research. 2010;**5**:21-27

[36] Zhang C, Yao J, Zhu C, Zhu M, Sun J. Soil heavy metal contamination and microbial ecological risk assessment—A case study. Chemical Engineering Transactions. 2016;**51**:307-312

[37] He CQ, Tan GE, Liang X, Du WYL, Zhi GY, Xhu Y. Effect of Zn-tolerant bacterial streins on growth and Zn accumulation in *Orychophragmus violaceus*. Applied Soil Ecology. 2010;**44**:1-5

[38] He LY, Zhang YF, Ma HY, Su LN, Chen ZJ, Wang QY, et al. Characterization of copper-resistant bacteria and assessment of bacterial communities in rhizosphere soils of copper-tolerant plants. Applied Soil Ecology. 2010;**44**:49-55

[39] Rutgers M, Wouterse M, Drost SM, Breure AM, Mulder C, Stone D, et al. Monitoring soil bacteria with community-level physiological profiles using biolog ECO-plates in the Netherlands and Europe. Applied Soil Ecology. 2016;**97**:23-35

[40] Liu G, Yu Y, Hou J, Xue W, Liu X, Liu Y, et al. An ecological risk assessment of heavy metal pollution of the agricultural ecosystem near a lead-acid battery factory. Ecological Indicators. 2014;**47**:210-218

[41] Chen J, He F, Zhang X, Sun X, Zheng J, Zheng J. Heavy metal pollution decreases microbial abundance, diversity and activity within particle-size fractions of a paddy soil. FEMS Microbiology Ecology. 2014;**87**:164-181

[42] Bansal OP, Singh G, Katiyar P. Effect of untreated sewage effluent irrigation on soil enzymatic activities. Journal of Environmental Biology. 2014;**35**: 641-647

[43] Bansal OP. Long term effects of organic manures (FYM, sewage sludge) amendments on soil enzymatic activities in an alluvial soil of Aligarh district: A 20 year study. Journal of Scientific Research in Allied Science. 2015;**1**:135-145

[44] Beyersmann D, Hartwig A. Carcinogenic metal compounds: Recent insights into molecular and cellular mechanisms. Archives of Toxicology. 2008;**82**:493-512

[45] Oliveira A, Pampulha ME. Effects of long-term heavy metal contamination on soil microbial characteristics. Journal of Bioscience and Bioengineering. 2006;**102**:157-161

[46] Meguro N, Kodama Y, Gallegos MT, Watanabe K. Molecular characterization of resistance-nodulation-division transporters from solvent- and drug-resistant bacteria in petroleum-contaminated soil. Applied and Environmental Microbiology. 2005;**71**:580-586

[47] Paul A, Wauters G, Paul AK. Nickel tolerance and accumulation by bacteria from rhizosphere of nickel hyperaccumulators in serpentine soil ecosystem of Andaman, India. Plant Soil. 2007;**293**:37-48

[48] Wyszkowska J, Borowik A, Kucharski M, Kucharski J. Effect of cadmium, copper and zinc on plants, soil organisms and soil enzymes. Journal of Elementology. 2013;**18**:769-796

[49] Nwuche CO, Ugoji EO. Effects of heavy metal pollution on the soil microbial activity. International Journal of Environmental Science & Technology. 2008;**5**:409-414

[50] Aoyama M, Tanaka R. Effects of heavy metal pollution of apple orchard surface soils associated with past use of metal-based pesticides on soil microbial biomass and microbial communities. Journal of Environmental Protection. 2013;**4**:27-36

[51] Wang T, Yuan Z, Yao J. A combined approach to evaluate activity and structure of soil microbial community in long-term potentially toxic elements contaminated soils. Environmental Engineering Research. 2018;**23**(1): 62-69

[52] Bansal OP, Mishra J. An estimation of microbial count including nitrogen fixing bacteria in agricultural fields of Aligarh district irrigated with untreated sewage water. International Journal of Science and Nature. 2012;**3**:259-262

[53] Kouchou A, Rais N, Elsass F, Duplay J, Fahli N, Ghachtouli EL. Effects of long term heavy metal contamination on soil microbial characteristics in calcareous agricultural lands (Saiss Plain North Morocco). Journal of Materials and Environmental Sciences. 2017;**8**:691-695

[54] Fengqiu A, Zhan D, Jialong L. Microbial diversity and community structure in agricultural soils suffering from 4 years of Pb contamination. Canadian Journal of Microbiology. 2018;**64**(5):305-316

[55] Hamsa N, Yogeshy GS, Koushik U, Patil L. Nitrogen transformation in soil: Effect of heavy metals. International Journal of Current Microbiology and Applied Sciences. 2017;**6**(5):816-832

[56] Mertens J, Wakelin SA, Broos K, Mclaughlin MJ. Extent of copper tolerance and consequences for functional stability of the ammonia-oxidizing community in long-term copper contaminated soils. Environmental Toxicology and Chemistry. 2010;**29**:27-37

[57] Ding Z, Wu J, You A, Huang B, Cao C. Effects of heavy metals on soil microbial community structure and diversity in the rice (*Oryza sativa* L. subsp. *japonica*, food crops Institute of Jiangsu Academy of Agricultural Sciences) rhizosphere. Soil Science and Plant Nutrition. 2017;**63**:75-83

[58] Nannipieri P, Ascher J, Ceccherini MT, Landi L, Pietramellara G, Renella G. Microbial diversity and soil functions. European Journal of Soil Science. 2003;**54**:655-670

[59] Kandeler E, Kampichler C, Horak O. Influence of heavy metals on the functional diversity of soil microbial communities. Biology and Fertility of Soils. 1996;**23**:299-306

[60] Frostegård Å, Tunlid A, Bååth E. Phospholipid fatty acid composition, biomass and activity of microbial communities from two soil types experimentally exposed to different potentially toxic elements. Applied and Environmental Microbiology. 1993;**59**:3605-3617

[61] Ellis RJ, Morgan P, Weightman AJ, Fry JC. Cultivation-dependent and independent approaches for determining bacterial diversity in heavy-metal-contaminated soil. Applied and Environmental Microbiology. 2003;**69**:3223-3230

[62] Khan S, Hesham AE, Qiao M, Rehman S, He J. Effect of Cd and Pb on soil microbial community structure and

activities. Environmental Science and Pollution Research. 2010;**17**:288-296

[63] Azarbad H, Nilinska M, Laskowski R, van Straalen NM, van Gestel CAM, Zhou J, et al. Microbial community composition and functions are resilient to metal pollution along two forest soil gradients. FEMS Microbiology Ecology. 2015;**91**:1-11

[64] Li S, Peng M, Liu Z, Shah SS. The role of soil microbes in promoting plant growth. Molecular Microbiology Research. 2017;**7**:30-37

[65] Tejada M, Gonzalef JL, Hernandez MT, Garcia C. Application of different organic amendments in a gasoline contaminated soil: Effect on soil microbial properties. Bioresource Technology. 2008;**99**:2872-2880

[66] Yu L, Cheng JM. Effect of heavy metals Cu, Cd, Pb and Zn on enzyme activity and microbial biomass carbon in brown soil. Advanced Materials Research. 2015;**1073-1076**:726-730

[67] Vig K, Megharaj M, Sethunathan N, Naidu R. Bioavailability and toxicity of cadmium to microorganisms and their activities in soil: A review. Advances in Environmental Research. 2003;**8**:121-135

[68] Nuaimi SA, Maktoom M. Effect of various potentially toxic elements on the enzymatic activity of bacterial alkaline phosphate [thesis]. United Arab Emirates University, Scholarworks@UAEU. 2010. p. 371

[69] Bhattacharyya P, Tripathy S, Kim K, Kim S. Arsenic fractions and enzyme activities in arsenic-contaminated soils by groundwater irrigation in West Bengal. Ecotoxicology and Environmental Safety. DOI: 10.1016/j.ecoenv.2007.08.015

[70] Ofoegbu CJ, Akubugwol EL, Dike CC, Madhka HCC, Ugwu CE, Obas NA. Effects of potentially toxic elements on soil enzymatic activities in the Ishiagu mining area of Ebonyl state—Nigeria. Forests. 2017;**8**:430-434

[71] Diana H, Goodey N, Mathieu C, Krumins JA, et al. Effect of metal contamination on microbial enzymatic activity in soil. Soil Biology and Biochemistry. 2015;**91**:291-297

[72] Waitrowska K, Komisarek J, Dluzewski P. Effects of heavy metals on the activity of dehydrogenases, phosphatases and urease in naturally and artificially contaminated soils. Journal of Elementology. 2015;**20**:743-756

[73] Derdzyan TH, Ghazaryan KA, Gevorgyan GA. The investigation of enzymatic activity in the soils under the impact of metallurgical industrial activity in Lori Marz, Armenia. International Journal of Environmental & Ecology Engg. 2015;**17**(5);439-442

[74] Gromakova N, Mandzhieva S, Minkina T, Sushkova S, et al. Effect of potentially toxic elements on the enzymatic activity of Haplic Chernozem under model experimental conditions. Journal of Biological Sciences. 2017;**17**(3):143-150

[75] Kandziora-Ciupa M, Ciepal R, Nadgorska-Socha A. Assessment of potentially toxic elements contamination and enzymatic activity in pine forest soils under different levels of anthropogenic stress. Polish Journal of Environmental Studies. 2016;**25**:1045-1051

[76] Welf G. Inhibitory effects of the total and water soluble concentrations of nine different metals on the dehydrogenase activity of a loess soil. Biology and Fertility of Soils. 1999;**30**:132-139

[77] Nowak J, Szymozak J, Slobodzian T. The test of qualification 50% threshold of toxicity of doses of different heavy metals for acid phosphatases.

Zeszyty Problemowe Postepow Nauk Rolniczych. 2003;**492**:241-248

[78] Li Q, Tang J, Wang T, Wu D, Jiao R, Ren X. Impacts of sewage irrigation on soil properties of farmland in China: A review. International Journal of Experimental Botany. 2018;**87**:1-21

[79] Dawki UM, Dikko AU, Noma SS, Aliu U. Heavy metals and physiochemical properties of soils in Kano urban agricultural lands. Nigerian Journal of Basic and Applied Sciences. 2013;**21**:239-246

[80] Rakesh Sharma MS, Raju NS. Correlation of heavy metal contamination with soil properties of industrial areas of Mysore, Karnataka, India by cluster analysis. International Research Journal of Environmental Sciences. 2013;**2**:22-27

[81] Tripathi A, Misra DR. A study of physiochemical properties and heavy metals in contaminated soils of municipal waste dumpsites at Allahabad, India. International Journal of Environmental Sciences. 2012;**2**:2024-2033

[82] Singare PU, Lokhande RS, Patha PP. Study on physico-chemical properties and heavy metal content of the soil samples from Thane Creek of Maharashtra, India. Interdisciplinary Environmental Review. 2010;**11**:38-56

[83] Nwaogu LA, Cosmas OU, Callistus II, Tobias NIE, Donatus CB. Effect of sublethal concentration of heavy metal contamination on soil physicochemical properties, catalase and dehydrogenase activities. International Journal of Biochemistry Research & Review. 2014;**4**(2):141-149

Chapter 3

Influence of Chemical Properties of Soil on the Corrosion Morphology of Carbon Steel Pipes

Muhammad Wasim and Shahrukh Shoaib

Abstract

Corrosive soils are responsible for the deterioration of buried underground utilities such as buried steel pipes. Frequent pipe failures are reported due to corrosive soil globally. Although soil's corrosion phenomenon has been understood and identified long time ago, pipe failures due to corrosive soil are uncontrollable and unavoidable despite the use of protective coatings and techniques such as cathodic protection. Therefore, it is essential to review the causes of soil's corrosivity for the protection of steel pipes. This chapter demonstrates the influence of varying moisture and chloride contents of soils on the corrosion of coated and uncoated steel pipes. Carbon steel specimens (coated and uncoated) were buried in soils of 20, 40, 60, and 80 wt.% moisture content, respectively, while the chloride concentration introduced in soil was 0, 5, and 10 wt.%, respectively. Through the analysis of experiments, it is revealed that the corrosion rate of pipes buried in soil increases with increase in moisture content up to critical moisture and chloride values. The influence of soil's moisture and chloride on the corrosion products formed on steel pipes was investigated and comprehensively explained in this chapter. Authors believe that the knowledge presented in this chapter can be applied to other structures or utilities buried in corrosive soils.

Keywords: soil, moisture content, chloride, steel pipes and corrosion products

1. Introduction

The influence of soil's chemical properties is reported as the root cause of failures of buried pipes [1–7]. The chemical constituents of soil react with the surface of unprotected buried pipes, which in turn results in the corrosion of pipes. However, there is still no complete preventive solution to the corrosion caused by the chemical constituents of soil even in the presence of advanced corrosion protection techniques.

As per above referred studies, soil's constituents cause corrosion of buried pipes; these include moisture contents, pH, temperature, soil resistivity, soil type, soil particle size, permeability, differential aeration, and sulphate-reducing bacteria (see references above). Researchers have adopted various approaches based on field testing (all above references) and experiments [8–10] to investigate these factors. Soil has been reported as the main stimulants causing failure of buried metallic pipes as shown in **Figure 1** [3].

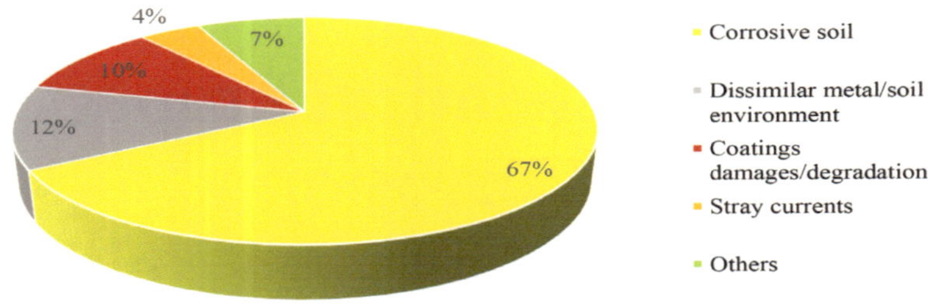

Figure 1.
Worldwide causes of corrosion of metallic buried pipes (redrawn, originally by [3]).

Among the various factors, the acidity (pH) and moisture contents of soil are stated as the most important key factors influencing corrosion of buried pipes as per the latest comprehensive reviews [5–7]. The influence of soil chemical properties such as pH on the buried pipes has been investigated by using the simulated soil solutions by many researchers to avoid the complex and heterogeneous soil structure by using varying quantities of acid [11–13]. Most studies on real soils are on the effect of moisture content and corresponding corrosion of buried pipes in laboratories [8–10].

The other notable chemical constitute of the soil, i.e., chloride, well known for its corrosion-causing capability particularly to reinforce concrete structures [14] can be responsible to the failure of steel pipes. However, from the comprehensive reviews, it can be found out that the research related to the effect of chlorides present in soil and the corresponding corrosion of carbon steel pipes is limited. Considering this gap, current research is conducted in which varying quantity of moisture and chlorides contents of the soil are taken into consideration for finding their effect and a coupled threshold value which would be useful to determine the service life of buried pipes.

2. Experimental methodology

The microstructure of corroded carbon steel samples exposed to different moisture and chloride conditions of soils was investigated. Authors conducted an experimental study on corrosion behaviour of carbon steel and zinc-electroplated and copper-electroplated carbon steels. Carbon steel specimens were exposed to 20, 40, 60, and 80 wt.% moisture, respectively, and the chloride concentration was kept at 0, 5, and 10 wt.%. First, the coupled effect of moisture and chloride which induced the maximum corrosion rate was evaluated. Then, zinc-electroplated and copper-electroplated steel specimens were exposed under similar aggressive coupled moisture and chloride condition. The details of the soils and specimen preparations and the chemical composition of the steel and soil used can be found elsewhere [15].

Experiments were performed under laboratory-controlled temperature of $27 \pm 1°C$. The set-up for the electrochemical measurements consisted of three electrode cells. The schematic diagram of the experimental test set-up is shown in **Figure 2**. In this figure, the specimen is represented by rectangular shape; was a carbon steel without coating, and with zinc- and copper-electroplated coatings, respectively; and was used as working electrode. In addition, two counter electrodes made of graphite and joined through electrical connector and copper/copper sulphate ($Cu/CuSO_4$) solution as reference electrode (RE) were used, respectively, for corrosion measurement of each specimen used in the current research. The surface area ratio between working electrode/counter electrode was kept at 0.909.

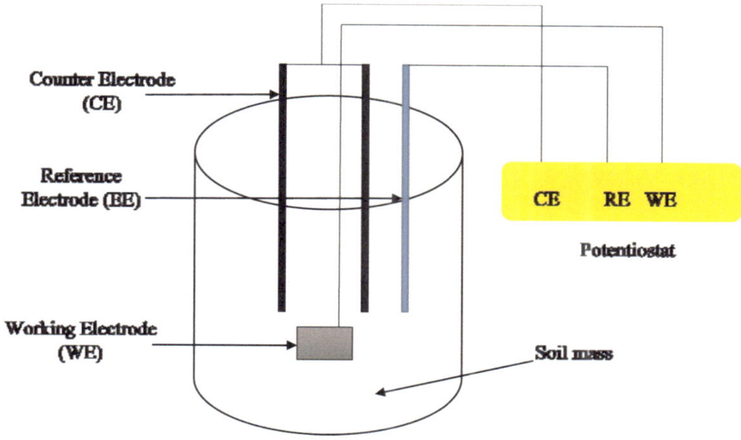

Figure 2.
Set-up for electrochemical measurement (modified after Shoaib et al. [15]).

Electrochemical measurements, i.e., electrochemical impedance spectroscopy (EIS), were performed using Autolab PGSTAT302N potentiostat/galvanostat. Before performing electrochemical measurement, metallic samples were cleaned as per ASTM G1-03. Then each specimen was buried into 1000 g of soil containing controlled moisture contents, chloride concentrations, and combination of both as discussed earlier. The open circuit potential of working electrode was observed for 3600 s. More details of the EIS and potentiodynamic polarisation procedures executed in the current research can be found in author's recent publication [15]. The EIS data was fit by using Nova 1.1.1 software for corrosion analysis. After knowing the corrosion rates of specimens, deep microstructure analysis of the corrosion morphology of all the specimens was executed using scanning electron microscopy (SEM). The corrosion results obtained and the related discussion are presented in the following section. More details about the preparation of specimens and soils can be found elsewhere [15].

3. Results and discussion

3.1 Electrochemical study on SS400 carbon steel

The electrochemical study is carried out to investigate the corrosion behaviour of metals. Shoaib et al. [15] examined the influence of moisture and chloride on the corrosion behaviour of SS400 carbon steel in the soil environment.

Electrochemical results showed that the corrosion rate of SS400 carbon steel sample increased with increase in moisture content up to 60 wt.% and decreased after this value. Moreover, with the addition of chloride, corrosion rate increased appreciably. The maximum corrosion rate was noticed for carbon steel exposed to soil containing 60 wt.% moisture and 5 wt.% moisture. Theoretically, carbon steel specimen buried in 60 wt.% moisture and 10 wt.% chloride should have more corrosion rate because of exposure of higher chloride content. Probably, the possible reason could be a non-homogeneous nature of the soil, or there is a possibility that organic contents might have caused the increase in soil's resistivity and hence the decrease in corrosion rate even in the presence of higher chlorides.

The corrosion rates of carbon steel samples under different exposure conditions are shown in **Figure 3**. The detailed discussion on the corrosion behaviour of various specimens can be found elsewhere [15]. After electrochemical

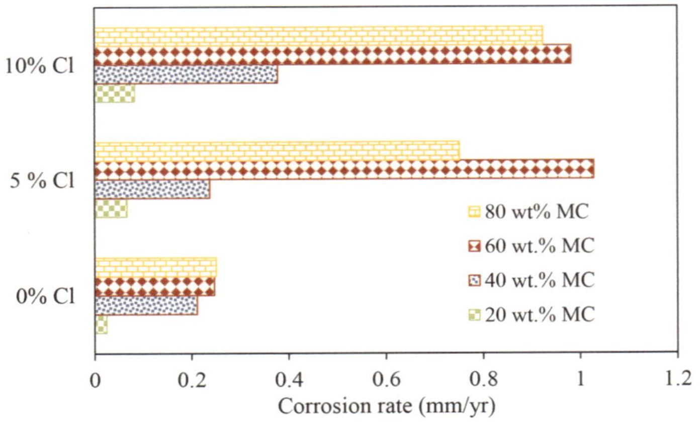

Figure 3.
Corrosion rates of SS400 steel samples under different moisture and chloride conditions (modified after Shoaib et al. [15]).

measurements of various steel samples, they were examined using laser microscope, scanning electron microscope (SEM), and energy-dispersive X-ray spectroscopy (EDS) to investigate the influence of chemicals in soils on the microstructure of buried steel samples. The discussion is presented in the following section.

3.2 Microscopic observation

First, the coupled effects of varying moisture and chloride on corrosion were measured [15]; then their subsequent impact on the microstructure of specimens was investigated using Olympus laser microscope. Corrosion patterns on specimens after exposure to various corrosive soils are shown in **Figures 4–6**; from these figures red rust can be seen on all samples. Interestingly, there was a further addition of red rust with the increase of chloride contents indicating the presence of iron oxides on the surface. Moreover, optical images also confirmed that the addition of chloride promoted corrosion progress. With further addition of chloride, more red rust was observed.

Two kinds of corrosion behaviour were observed in specimens, i.e., general and localised corrosion. Carbon steel sample exposed to soil containing 60 wt.% moisture and 5 wt.% chloride showed localised corrosion, while the samples buried in soil of 80 wt.% moisture and 10 wt.% chloride suffered general corrosion. **Figure 7** shows the corrosion morphology of copper-electroplated and zinc-electroplated steel samples. However, for copper-electroplated steel samples, there was no clear pattern of corrosion. On the other hand, zinc-electroplated steel indicated localised corrosion.

3.3 Corrosion product morphology

SEM analyses were performed after the samples were corroded to various soil conditions. **Figure 8** shows the SEM micrographs of low and high magnifications of carbon steel. From this figure, a porous and honeycomb-like structure appeared, which is also reported by earlier researchers [16]. **Figure 9** shows EDS spectra of elements present in corrosion product of carbon steel, while the elements in the corrosion layer of carbon steel are shown in **Table 1**. The presence of sodium (Na) on a metallic surface, indicating that cation in soil penetrated through corrosion product layer and reached the sample's surface. As a result of this penetration, the corrosion process is accelerated. However, it has been reported in literature that the

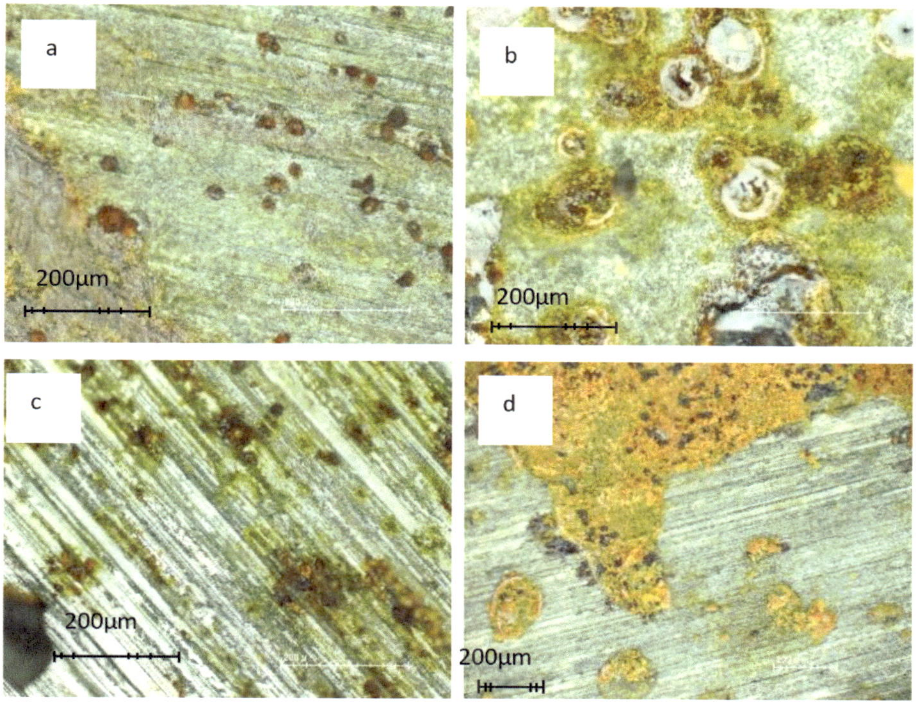

Figure 4.
Corrosion morphology of carbon steels tested in different soil moisture contents (a = 20 wt.% MC; b = 40 wt.% MC; c = 60 wt.% MC; d = 80 wt.% MC).

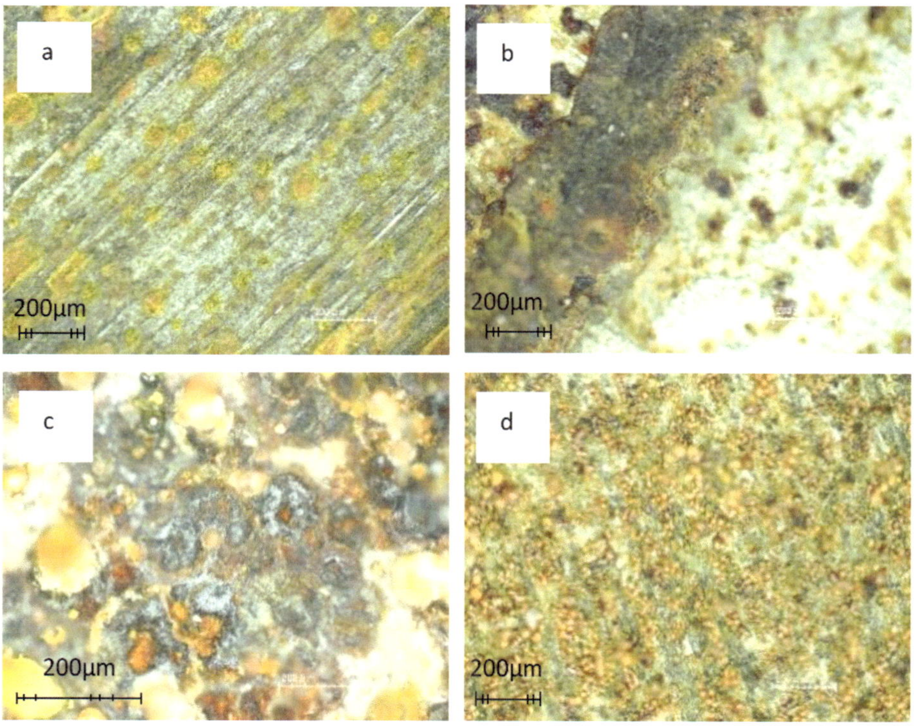

Figure 5.
Corrosion morphology of carbon steels tested in soil with 5 wt.% chloride a = 20 wt.% MC; b = 40 wt.% MC; c = 60 wt.% MC; d = 80 wt.% MC).

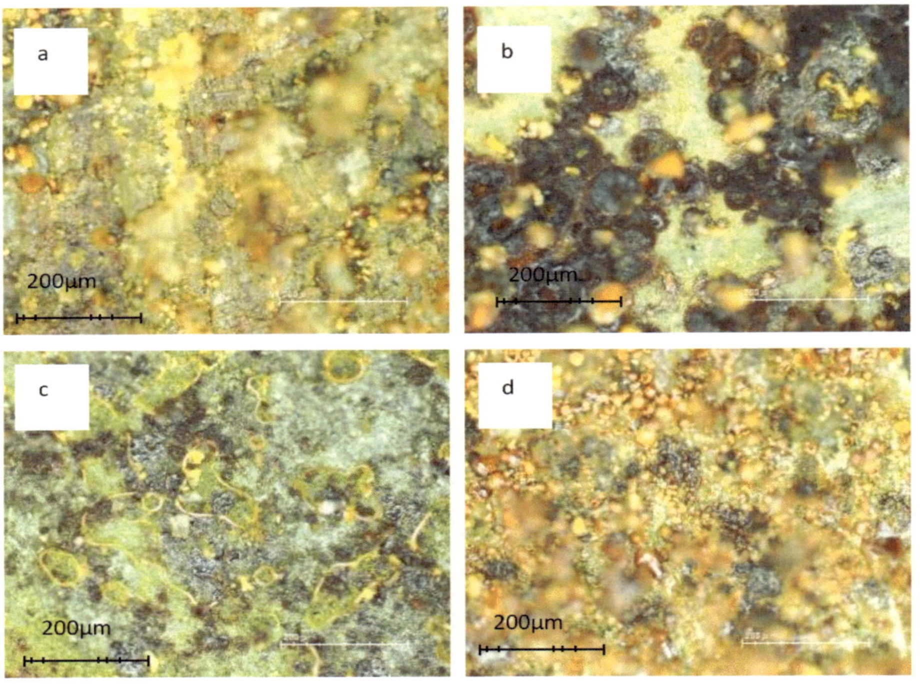

Figure 6.
Corrosion morphology of carbon steels tested in soil with 10 wt.% chloride (a = 20 wt.% MC; b = 40 wt.% MC; c = 60 wt.% MC; d = 80 wt.% MC).

initial corrosion product formed on carbon steel is α-FeOOH which provides a shield to substrate metal against corrosion [17, 18].

Figure 10 SEM images of zinc-electroplated steel with a low and high magnification. A crystalline structure was observed which spread over the surface area of the sample. This indicates the occurrence of generalised corrosion mechanism [19]. The EDS spectra of elements present in corrosion product of zinc-electroplated steel are shown in **Figure 11**. Three locations were selected for the determination of elements in corrosion product. Elements obtained from the corroded layer of zinc-coated steel are shown in **Table 2**. Elements determined were Zn, Fe, O, and Cl

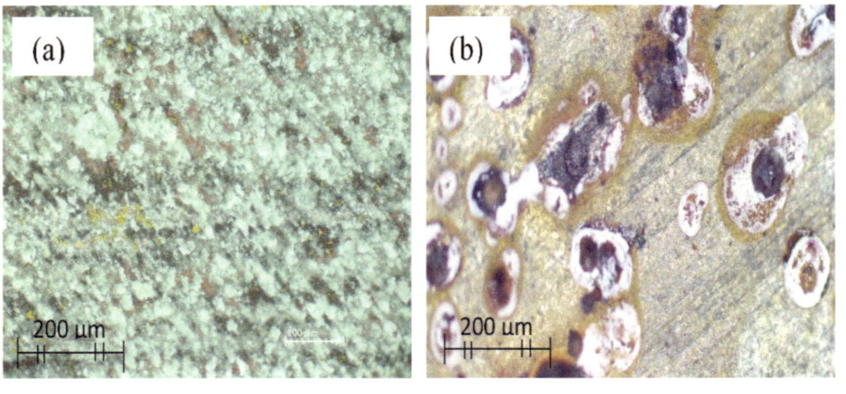

Figure 7.
Corrosion morphology of (a) copper-electroplated and (b) zinc-electroplated steels tested in soil with 60 wt.% moisture and 5 wt.% chloride.

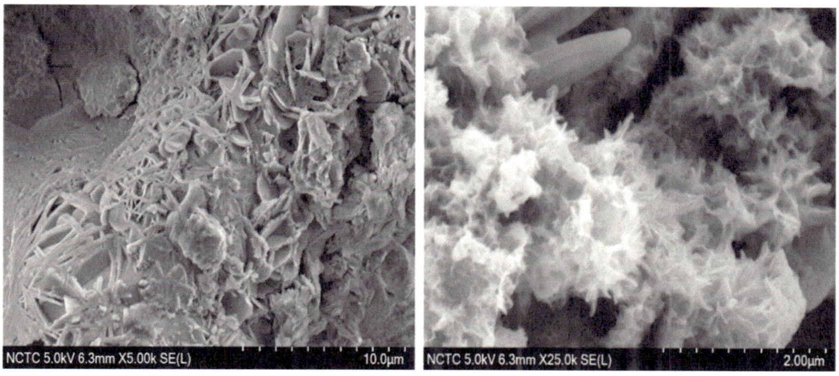

Figure 8.
SEM images of carbon steel buried in soil of 60 wt.% moisture and 5 wt.% chloride.

which indicated the likelihood of zinc oxide (ZnO) is formed. Zinc oxide provides a protective layer and protects the substrate metal from further corrosion [20]. However, the presence of Fe indicates that protective layer of zinc was removed partially due to soil aggressiveness.

Figure 12 shows SEM micrographs of copper-coated steel with a low and high magnification. From this figure corrosion pattern of granular and compact structure can be seen which is also reported in the literature [21]. However, **Figure 13** demonstrates EDS spectra of elements present in corrosion product of copper-coated steel, also listed in **Table 3**. From the table it can be seen that Br was present in trace

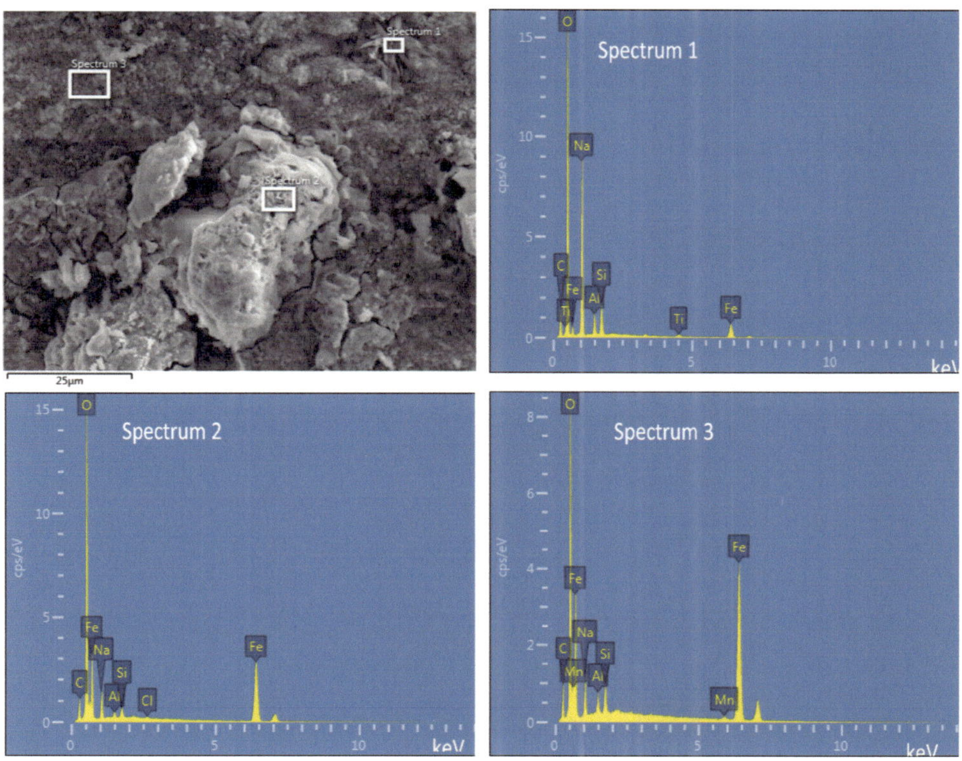

Figure 9.
EDS spectra of carbon steel tested in soil with 60 wt.% moisture and 5 wt.% chloride.

Element	Spectrum 1	Spectrum 2	Spectrum 3
Fe	9.5	40.1	56.6
O	51	43.7	26.5
Na	20.4	4.1	3.6
C	13.7	10.6	10.1
Si	2.8	0.8	1.4
Al	2	0.6	1
Mn	—	—	0.8
Cl	—	0.2	—
Ti	0.6	—	—

Table 1.
*EDS analyses (wt.%) of carbon steel of selected spots in **Figure 8**.*

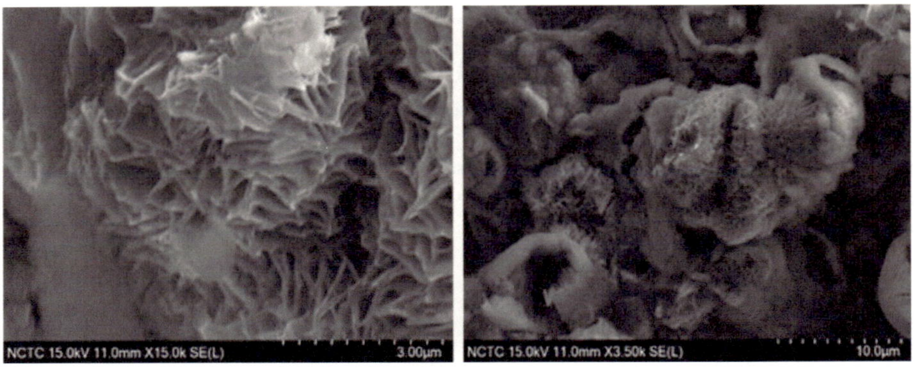

Figure 10.
SEM images of zinc-coated steel tested in soil with 60 wt.% moisture and 5 wt.% chloride.

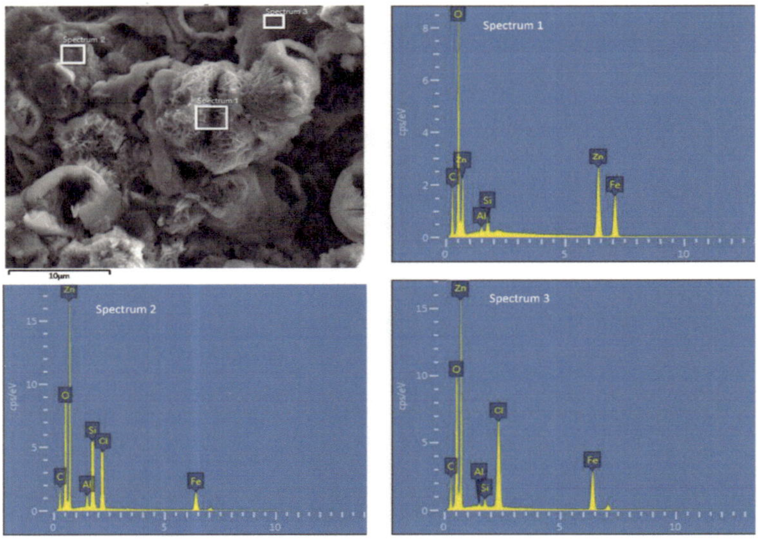

Figure 11.
EDS spectra of zinc-coated steel tested in soil with 60 wt.% moisture and 5 wt.% chloride.

Element	Spectrum 1	Spectrum 2	Spectrum 3
Zn	43.4	31.2	26.2
O	22.4	17.6	30.7
Fe	12.1	12.7	9.3
C	20.4	13.8	12.6
Si	1.2	8.6	4.2
Al	0.5	0.9	0.3
Cl	—	15.2	16.7

Table 2.
EDS analyses (wt.%) of zinc-coated steel for selected spots in ***Figure 9****.*

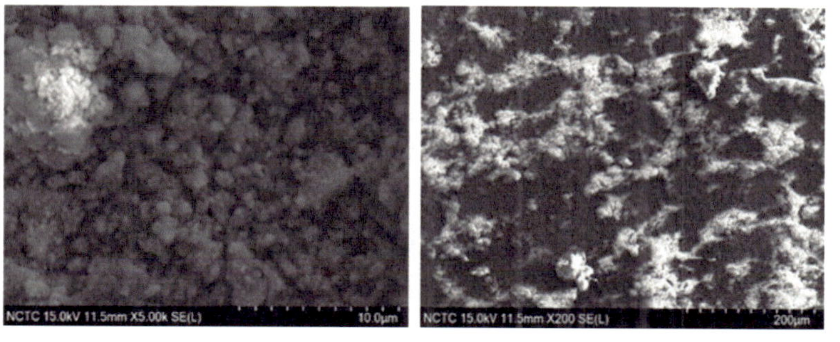

Figure 12.
SEM images of copper-coated steel tested in soil with 60 wt.% moisture and 5 wt.% chloride.

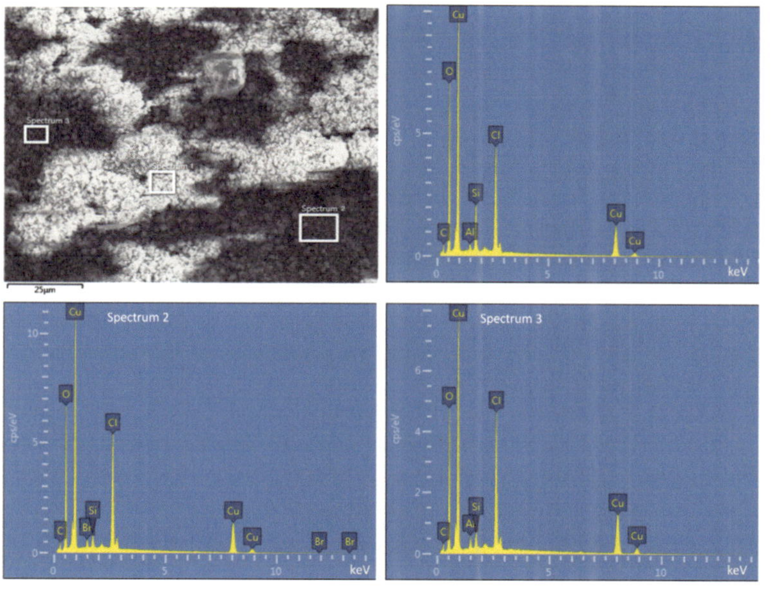

Figure 13.
EDS spectra of copper-coated steel tested in soil with 60 wt.% moisture and 5 wt.% chloride.

amount in substrate metal. Br contributes to the acceleration of corrosion, but it is less aggressive than Cl. From the analysis results, it is evident that there is less quantity of Br (1.6%) in corrosion product of copper-coated steel which means Br did

Element	Spectrum 1	Spectrum 2	Spectrum 3
Cu	45.2	47.2	46.3
O	29.8	27.5	27.3
Cl	11.7	13.5	15.3
C	8.7	9	8.7
Si	4	1.6	1.4
Al	0.6	1.1	1.1

Table 3.
EDS analyses (wt.%) of copper-coated steel for selected spots in ***Figure 12****.*

not contribute to corrosion significantly. The elements present in the corrosion product show the possibility of the presence of copper chloride and copper oxide [21, 22].

The presence of Cl element in the corrosion layer of carbon, zinc-coated, and copper-coated steels was obviously due to the addition of NaCl. The Cl anion is classified as aggressive because it directly contributes in electrochemical reaction causing corrosion. Cl anion present in coated samples reveals that it has a strong tendency to promote corrosion rate even if the metallic surface is coated with zinc or copper.

3.4 Energy-dispersive X-ray spectroscopy (EDS) mapping

Cross-sectional EDS map of specimens buried in soil containing 80 wt.% moisture is shown in **Figure 14a** and **b**. In **Figure 14a**, O-K and Fe-K element maps show that oxygen and iron are uniformly distributed. The co-existence of O and Fe elements demonstrates that oxides of iron are present in corrosion products. Generally, α-FeOOH and γ-FeOOH are observed as corrosion products of steel buried in the soil environment. A C-K element from epoxy resin has a non-uniform distribution. At site 2 (**Figure 14b**), there is an excessive concentration of C-K due to the epoxy resin, and there is less concentration of O-K and O-K at this site.

Figure 15a and **b** demonstrates cross-sectional EDS map of carbon steel sample exposed to soil containing 60 wt.% moisture and 5 wt.% chloride at site 1. A pit can be observed on carbon steel sample. It confirms that exposure of carbon steel to soil containing chloride accelerates the corrosion; as a result pitting is observed. The layer of oxygen has a variable thickness, and C layer from epoxy has a non-uniform thickness due to which no chloride contents were observed. The possible reason is that any element having a concentration less than 1% cannot be mapped by EDS mapping technique because peaks of elements having less concentration are difficult to separate from the background. **Figure 15b** shows EDS mapping at site 2, where the pit is less wide than site 1.

Figure 16a and **b** illustrates the EDS mapping of steel specimens buried in soil containing 60 wt.% moisture and 10 wt.% chloride. This specimen showed entirely different behaviour as there was less O-K concentration, which also confirms the SEM results of no stable oxide layer, while the concentration of Fe-K came out to be high. It can be interpreted from mapping results that there are fewer oxides present in this condition.

Theoretically, carbon steel specimens exposed to 10 wt.% chloride should be more corroded than 5 wt.% chloride; however, experimentally this was not observed. The reason for this phenomenon is probably due to the fact that soil is a non-homogeneous, and its properties vary within the soil itself. There is a possibility that soil sample used for 60 and 5 wt.% chloride conditions might have considerable chloride already present in it. Furthermore, it is also possible that the soil sample used

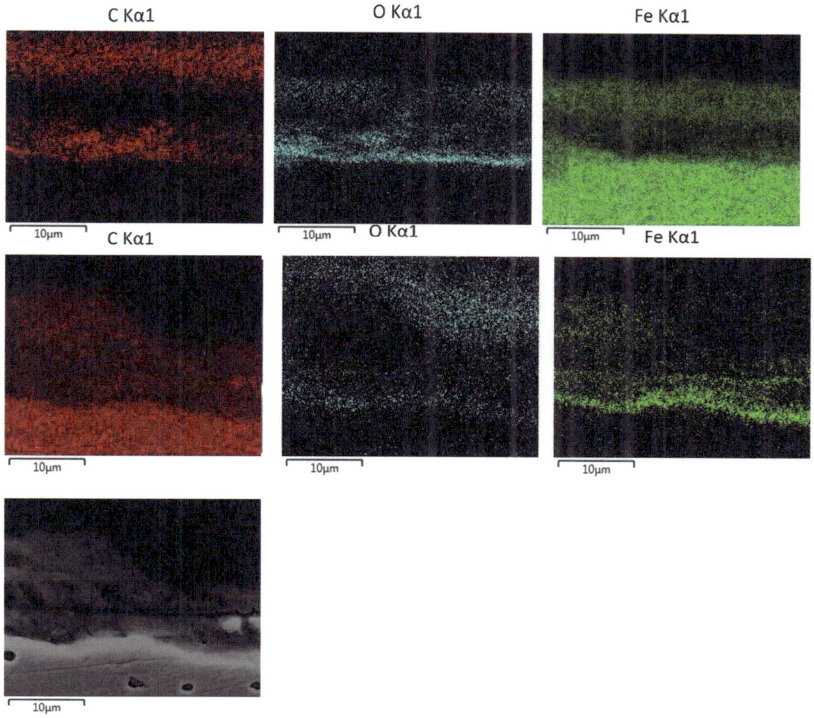

(a)

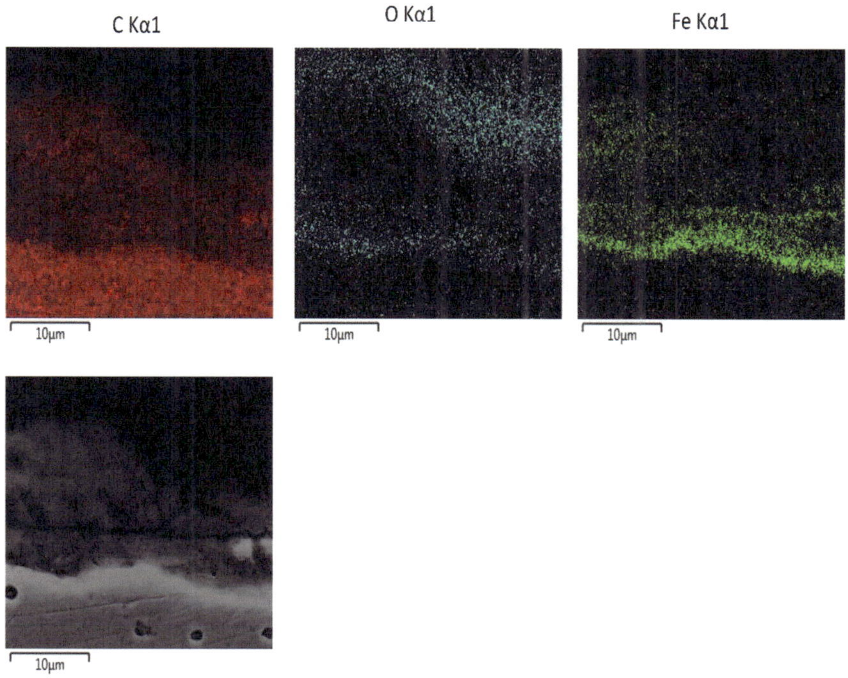

(b)

Figure 14.
(a) EDS map of carbon steel exposed to soil containing 80 wt.% moisture (site 1). (b) EDS map of carbon steel exposed to soil containing 80 wt.% moisture (site 2).

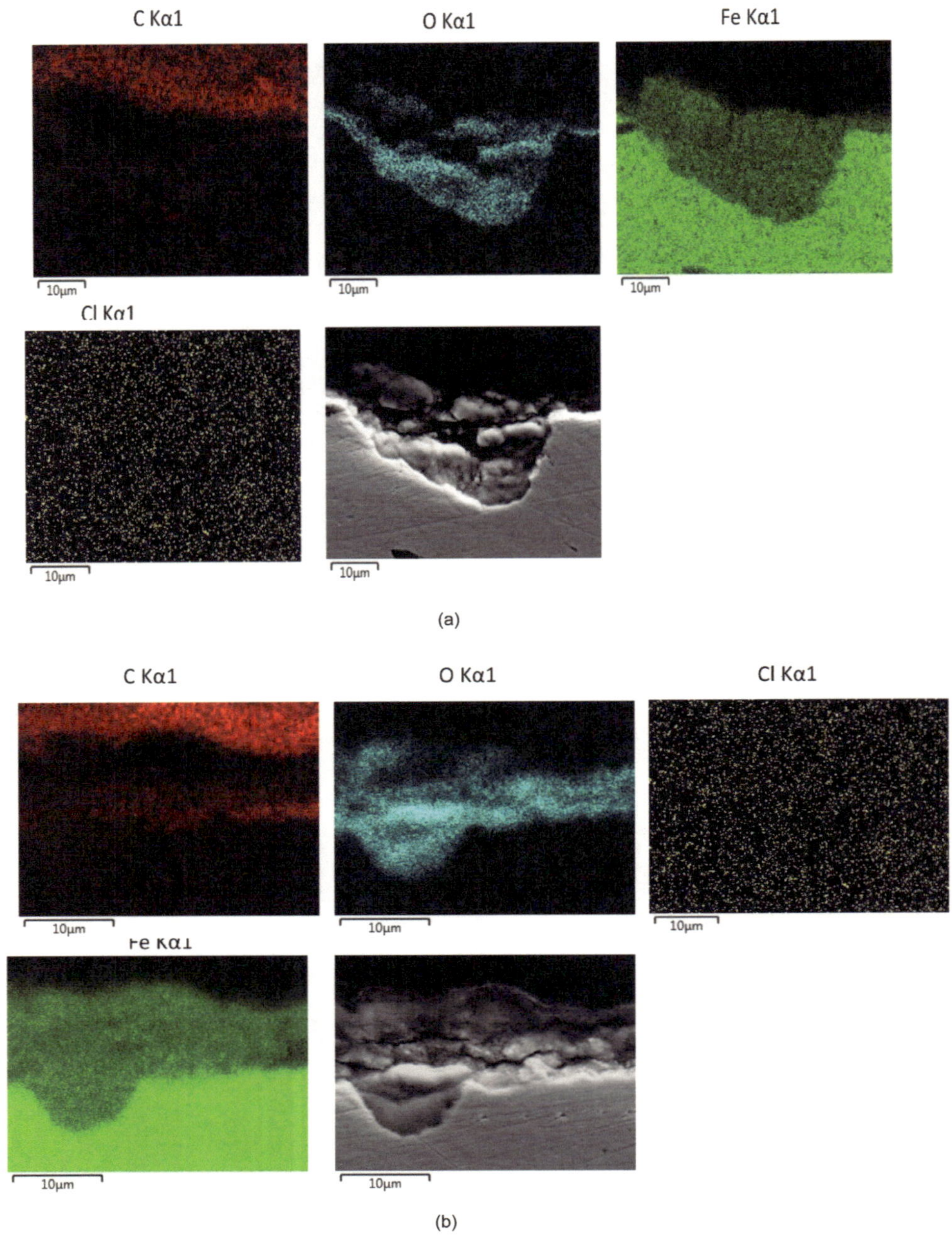

Figure 15.
(a) EDS map of carbon steel exposed to soil containing 60 wt.% moisture and 5 wt.% chloride (site 1). (b) EDS map of carbon steel buried in soil containing 60 wt.% moisture and 5 wt.% chloride (site 2).

for 60 and 10 wt.% might have organic contents. The presence of organic contents in soil increases the soil resistivity and ultimately decreases the corrosion rate. It is concluded from the above discussion that the addition of chloride in soil accelerates the corrosion rate significantly and in a short time. The increase in exposure duration to chloride-contaminated soil can lead to pitting of carbon steel.

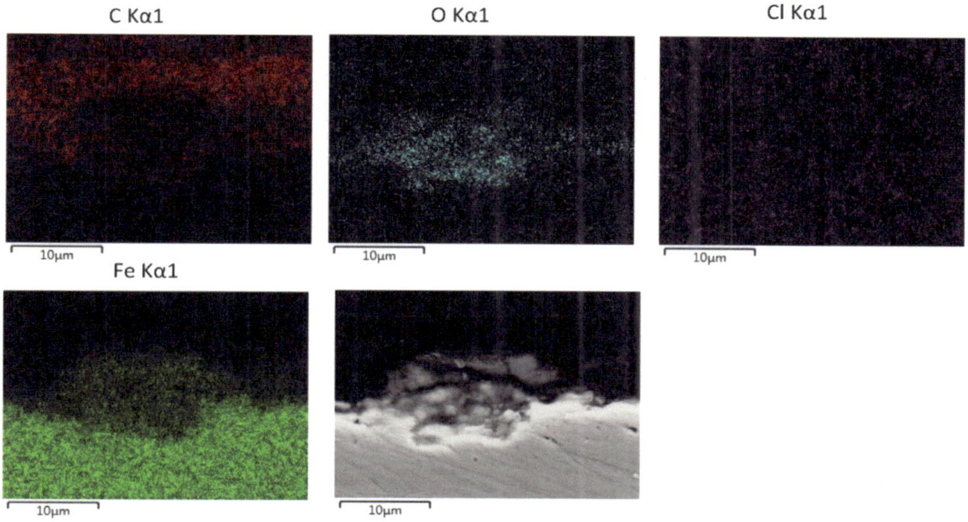

(a)

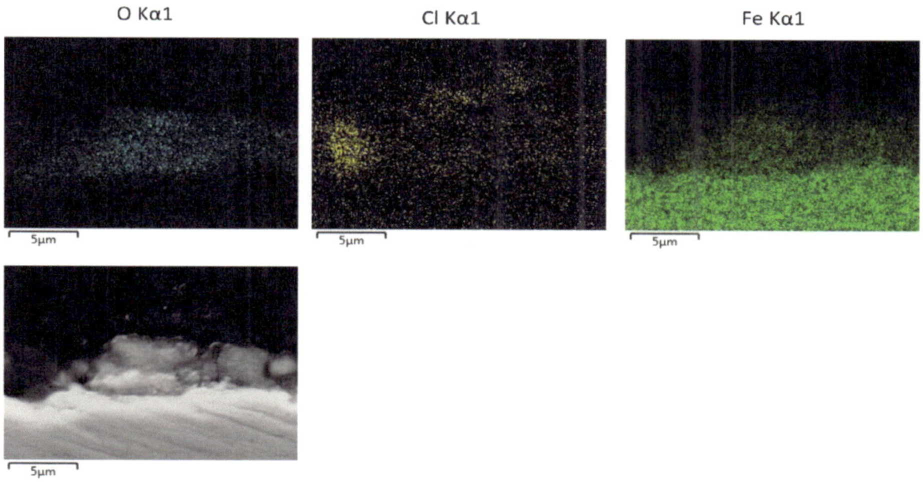

(b)

Figure 16.
(a) EDS map of carbon steel exposed to soil containing 60 wt.% moisture and 10 wt.% chloride (site 1).
(b) EDS map of carbon steel exposed to soil containing 60 wt.% moisture and 10 wt.% chloride (site 2).

4. Conclusions

In this chapter, a study related to the influence of soil's varying moisture and chloride contents on the corrosion and subsequent microstructure of coated and uncoated carbon steel pipes is presented. From the experimental findings of corrosion and then after carrying out microstructural analysis, a threshold value for moisture and chloride contents is determined beyond which no further addition of chloride and moisture contents can cause corrosion and deterioration of microstructure of carbon steel. The results presented in this paper have practical application for the protection of coated and uncoated carbon steel pipes in soils. This study can help owners of the steel pipes to decide which type of coating to be used for the protection of the carbon steel pipes in aggressive soil conditions such as those presented in this paper. The current research is further extended for longer exposures and evaluating the influence of corrosion on the mechanical properties of buried steel pipes.

Author details

Muhammad Wasim[1]* and Shahrukh Shoaib[2]

1 RMIT University, Melbourne University, Melbourne, Australia

2 United Arab Emirates University, Al Ain, United Arab Emirates

*Address all correspondence to: wasim_oct@hotmail.com

IntechOpen

© 2019 The Author(s). Licensee IntechOpen. This chapter is distributed under the terms of the Creative Commons Attribution License (http://creativecommons.org/licenses/by/3.0), which permits unrestricted use, distribution, and reproduction in any medium, provided the original work is properly cited.

References

[1] Romanoff. Underground Corrosion. Superintendent of Documents Washington, D.C. National Bureau of Standards Circular; 1957:579

[2] Kleiner Y, Rajani B. Comprehensive review of structural deterioration of water mains: Statistical models. Urban Water. 2001;**3**:131-150. DOI: 10.1016/S1462-0758(01)00033-4

[3] Romer AE, Bell GE. Causes of external corrosion on buried water mains. In: Pipelines: Advances in Pipelines Engineering & Construction; 2001. pp. 1-9. DOI: 10.1061/40574(2001)20

[4] Alamilla JL, Espinosa-Medina MA, Sosa E. Modelling steel corrosion damage in soil environment. Corrosion Science. 2009;**51**:2628-2638. DOI: 10.1016/j.corsci.2009.06.052

[5] Cole I, Marney D. The science of pipe corrosion: A review of the literature on the corrosion of ferrous metals in soils. Corrosion Science. 2012;**56**:5-16. DOI: 10.1016/j.corsci.2011.12.001

[6] Asadi ZS, Melchers RE. Long-term external pitting and corrosion of buried cast iron water pipes. Corrosion Engineering, Science and Technology. 17 Feb 2018;**53**(2):93-101

[7] Wasim M, Shoaib S, Mubarak NM, Asiri AM. Factors influencing corrosion of metal pipes in soils. Environmental Chemistry Letters. 2018;**1**:1-9

[8] Qin F, Jiang C, Cui X, Wang Q, Wang J, Huang R, et al. Effect of soil moisture content on corrosion behavior of X70 steel. International Journal of Electrochemical Science. 2018;**13**:1603-1613. DOI: 10.20964/2018.02.32

[9] Gupta S, Gupta B. The critical soil moisture content in the underground corrosion of mild steel. Corrosion Science. 1979;**19**:171-178. DOI: 10.1016/0010-938X(79)90015-5

[10] Wan Y, Ding L, Wang X, LI Y, Sun H, Wang Q. Corrosion behaviors of Q235 steel in indoor soil. International Journal of Electrochemical Science. 2013;**8**:12531-12542

[11] Noor EA, Al-Moubaraki AH. Influence of soil moisture content on the corrosion behavior of X60 steel in different soils. Arabian Journal for Science and Engineering. 1 Jul 2014;**39**(7):5421-5435

[12] Liu ZY, Li XG, Du CW, Lu L, Zhang YR, Cheng YF. Effect of inclusions on initiation of stress corrosion cracks in X70 pipeline steel in an acidic soil environment. Corrosion Science. 2009;**51**:895-900

[13] Yang Y, Cheng F. Effect of stress on corrosion at crack tip on pipeline steel in a near-neutral ph solution. Journal of Materials Engineering and Performance. 2016;**25**:4988-4995. DOI: 10.1007/s11665-016-2369-9

[14] Zhou Y, Gencturk B, Willam K, Attar A. Carbonation-induced and chloride-induced corrosion in reinforced concrete structures. Journal of Materials in Civil Engineering. 2014;**27**(9):04014245

[15] Shoaib S, Srinophakun TR, Palsson NS. Influence of soil conditions on corrosion behavior of buried coated and uncoated carbon steels. In: IEEE International Conference on Innovative Research and Development (ICIRD); 11-12 May 2018. DOI: 10.1109/ICIRD.2018.8376310

[16] Song Y, Jiang G, Chen Y, Zhao P, Tian Y. Effects of chloride ions on corrosion of ductile iron and carbon steel in soil environments. Scientific Reports. 2017;**7**:1-13. DOI: 10.1038/s41598-017-07245-1

[17] Nie X, Li X, Du C, Huang Y, Du H. Characterization of corrosion products formed on the surface of carbon steel by Raman spectroscopy. Journal of Raman Spectroscopy. 2009b;**40**:76-79. DOI: 10.1002/jrs.2082

[18] Kwon S, Shinoda K, Suzuki S, Waseda Y. Influence of silicon on local structure and morphology of c-FeOOH and a-FeOOH particles. Corrosion Science. 2007;**49**:1513-1526. DOI: 10.1016/j.corsci.2006.07.004

[19] Lima-Neto PD, Adriana NC, Colares PR, Walney SA. Corrosion study of electrodeposited Zn and Zn-Co coatings in chloride medium. Journal of the Brazilian Chemical Society. 2007;**18**: 1164-1175. DOI: 10.1590/S0103-50532007000600010

[20] Spathis P, Poulos I. The corrosion and photocorrosion of zinc and zinc coatings. Corrosion Science. 1995;**37**: 673-680

[21] Kamboj A, Paghupathy Y, Rekha MY, Srivastavai C. Morphology, texture and corrosion behavior of nanocrystalline copper–graphene composite coatings. Journal of Materials. 2017;**69**:1149-1154

[22] Afonso FS, Neto MMM, Mendonça MH, Pimenta G, Proença L, Fonseca ITE. Copper corrosion in soil: Influence of chloride contents, aeration and humidity. Journal of Solid State Electrochemistry. 2009;**13**:1757-1765. DOI: 10.1007/s10008-009-0868-4

Chapter 4

Radioactive Isotopes in Soils and Their Impact on Plant Growth

Jelena Markovic and Svetlana Stevovic

Abstract

In 1999, Serbia was bombarded by NATO. One of the cities most affected by the consequences of bombardment with uranium is the city of Vranje, where the consequences are felt even today. Due to the influence of uranium, the mortality rate has increased. This paper presents the effects of some of the radionuclides that have contaminated the soil, as well as the connection between soil and plants that grow on that soil. The performed measurements of radionuclides (^{226}Ra, ^{40}K, ^{232}Th, ^{238}U, and ^{235}U). The results show that the content of each of these radionuclides has different concentrations, but what is important is that some values are even below the detection limit, corn <0.06 ^{235}U on the location Korbevac and wheat <0.04 ^{235}U on the location Bujkovac. On the three and all of these gated locations, the calculated values of the transfer factors for ^{40}K were in the range of 0.144–0.392, while in the case of ^{226}Ra, the transfer factors ranged from 0.008 to 0.074. Only one value (0.051) was obtained for the transfer factor of ^{232}Th. Specific activities of ^{137}Cs, as well as uranium isotopes, in all the investigated cereal samples, were below minimal detectable activity concentrations. The ratio of radionuclides in soil and plants is of great importance for human nutrition.

Keywords: soil, plants, radioactive isotopes, monitoring, mortality

1. Introduction

Natural radioactivity in the environment, originating from the naturally occurring radionuclides of ^{232}Th, ^{238}U, and ^{235}U radioactive series and ^{40}K, largely contributes to the natural irradiation of man and biota, which can be external and/or internal (ingestion and inhalation). Natural radionuclides land characteristic of α and β radioactive decay [1]. The biggest number of radionuclides belongs to a radioactive series, which naturally has three. These three series start as radioisotope, so-called parent: ^{238}U (series of 4n + 2), ^{235}U (series of 4n + 3), and ^{232}Th (string 4n). A series of successive radioactive decays occur from parents whose offspring core is also unstable and is subject to decay. The process of disintegration ends stable isotopes, and for those strings to the ^{206}Pb, ^{207}Pb, and ^{208}Pb, respectively, from the radionuclide which does not belong to any of the radioactive series, the most important is the soil radionuclide ^{40}K. Gamma radiation created during the radioactive decay of uranium and thorium series, as well as ^{40}K, largely contributes to the natural irradiation of alive council (man and biota), which can be external and/or internal (ingestion and inhalation) [2]. The concentration of natural radionuclides depends on the composition of the soil. According to the

IntechOpen

report of the UNSCEAR, the medium activity concentration of ^{238}U, ^{232}Th, and ^{40}K in the soil in the world amounts 33.45 and 412 mCi kg-1, respectively. The ranges of concentrations of ^{238}U, ^{232}Th, and ^{40}K in the soil in Europe are 2–330, 2–190, and 40–1650 Bq kg^{-1} [3].

Besides the natural radionuclides, due to various human activities, different manmade radionuclides entered the environment. The most significant among them is ^{137}Cs ($T_{1/2}$ = 30y), found in the environment mostly as a result of the nuclear tests in the 1960s ties and the Chernobyl nuclear plant accident in 1986 [4]. ^{137}Cs is bound in the surface layers of soil and is washed out and redistributed in the ecosystem for a longer period of time due to its long half-life. Thus, only a small amount of it is present in plants today. It is well-known that ^{137}Cs isotopes take important part in the environment, due to their good assimilation by plants, which are used to feed the animals and finally human beings [5]. The reported values of ^{137}Cs in the agricultural soil in the north part of Serbia, on several locations near the city of Novi Sad, are in the range of 1.5–12.6 Bq/kg [6].

Soil is a complex material composed of mineral (inorganic) as well as an organic matter that originated from plant decomposition. It is a compact matter providing necessary micro- and macro-nutrition elements for plants to function and grow. Cereals as wheat, corn and barley are important component of everyday human diet [21]. Most of the radionuclide cereals are absorbed from soil, so the values of transfer factors are important in the studies of the transport and distribution of radionuclides in the "soil–plant–animal-human" chain, as well as in the evaluation of the radiation risk [7]. Transfer factors (TF) are crucial in the radionuclide transport models, in the environment as well as in the evaluation of the level of the specific activities of radionuclides in agricultural crops [8]. The main factors determining the level of TF are radionuclide itself, type of plant, type (physical and chemical characteristics) of soil, concentrations of stable chemical elements in soils [9], as well the local climate [10].The values of transfer factors should provide the basis for theoretical analysis on the different uptakes of elements not involved in physiological and biochemical processes in plants [11].

The main goal of this paper was to investigate transfer factor, because it can give crucial information about the possible quantity of radioactive and other

Figure 1.
Pcinja District.

toxic materials endangered for human's health. Starting from the soil, through the plants, they enter in the food chain and consequently they reach in the human body and affect mortality. The transport processes in the "soil–plant" systems for radionuclides (^{226}Ra,^{40}K, ^{232}Th, ^{238}U, ^{235}U) and ^{137}Cs in the Pcinja District. Pcinja District is located in the southern part of the Republic of Serbia. It covers the city of Vranje and the municipalities Vladicin Han, Surdulica, Bosilegrad, Trgoviste, Bujanovac, and Presevo (**Figure 1**). The Pcinja District has not been investigated yet.

2. Methodology

The samples of soils and cereals were collected in 2014, in the area of the city of Vranje, on three locations: the villages of Bujkovac, Korbevac, and Suvi Dol. The type of soils was the same on all the locations (the so-called *gajnjaca*). *Gajnjaca* belongs to well-drained soils, its chemical characteristics depending on the level of utilization, degree of erosion, chemical characteristics of the main substrate, and level of development. The content of humus in the gajnjaca soils is in the range of 2–5%. This type of soil in neutral or low acetous has a high capacity of adsorption, and the dominant ions in it are Ca and Mg. Its color is brown, reddish, or red depending of the content of aluminum and iron. It is very suitable for farming, wine growing, and afforestation.

2.1 Sampling sites and sample collection

About 11 samples of cultivated and uncultivated soils and 7 samples of cereals were collected. Sampling sites coordinates are present in **Table 1**. The samples of soils were taken from different depths that also differ from one to another sampling site. The depths that the soils were sampled from were 0–5 cm, 0–10 cm, and 0–20 cm on the sites Korbevac and Bujkovac and 0–5 cm, 5–10 cm, and 10–15 cm at the sampling site Suvi Dol. The samples of grain were taken from plots where the soil samples were taken. First, soil samples were taken for testing, on the land-sown cereals that are taken when they are ripe for examination. The position of the sampling sites is presented in **Figure 2**.

The radioactivity of the samples of soils was determined by gamma spectrometry in the Institute for Nuclear Sciences "Vinca" in the Laboratory for Radiation and Environmental Protection.

The mass of cereal and soil samples for analysis is necessarily 1 kg. The samples of soils were cleaned of mechanical impurities, stones and plant material, and dried at 105°C for 24 h. Samples of cereals were dried at room temperature and mineralized at 450°C. Soils were measured in Marinelli geometry (volume 500 ml) and cereals in cylinder bottles (volume 125 ml). The radioactive equilibrium was

Site	Coordinates		Elevation (m)	Sampling date
	(North latitude)	(East longitude)		
Bujkovac	42°33′26″	22°00′35″	718	09.11.2014.
Korbevac	42°23′06″	21°44′24″	441	05.11.2014.
Suvi Dol	42°33′07″	21°56′05″	359	11.11.2014.

Table 1.
Coordinates of the location of soil samples.

Figure 2.
Sampling sites on the territory of the city of Vranje.

achieved in all the samples, as they have been sealed by bee wax and left for 30 days before measuring. The samples of grain were taken at the stage of full maturity of the technology and to hand. Samples of cereals (fruit cereal) were dried in air at room temperature for at least 3 weeks and then were crushed and mineralized at a temperature of 450°C for 24 hours, dry-ashing method [12].

The specific activity of natural ^{226}Ra was determined by analyzing the spectra of its daughters [13], ^{214}Pb and ^{214}Bi, at the energies of 295, 352, 609, 1120, and 1764 keV [14]. Radionuclide ^{232}Th was determined by its daughter ^{228}Ac at the energies of 338 and 911 keV. The activities of ^{40}K and ^{137}Cs were determined at the energies of 1460 and 661.6 keV, respectively. The activity of ^{235}U and ^{238}U is determined by establishing a radiochemical equilibrium between ^{226}Ra and ^{214}Bi using photo-peak at energies around 186 keV-a [15, 16].

2.2 Standard gamma spectrometry

The gamma spectrometry was performed on three HPGe detectors (CANBERRA) with relative efficiencies of 18, 20, and 50%; the resolution of all of the detectors was 1.8 keV at 1332 keV. For the samples of soils, the detectors were calibrated by a reference radioactive material—a silicone resin matrix, Czech Metrological Institute, Praha, 9031-OL-420/12, total activity 41.48 kBq on 31.08.2012 (^{241}Am, ^{109}Cd, ^{139}Ce, ^{57}Co, ^{60}Co, ^{203}Hg, ^{88}Y, ^{113}Sn, ^{85}Sr ^{137}Cs). The gamma-spectrometric measurements of radioactivity in soil samples was used and the ultra-low-background germanium detector-type GMX (extended energy range from 10 keV to 3 MeV-a manufacturer of ORTEC, the nominal efficiency of 32% in passive and active protection). Passive safety lead is made up of a thickness of 12 cm in the form of a cylinder and coated with a layer of tin and copper. The active protection (veto detectors) is the five plastic scintillation detectors which are anticoincidence mode working with HPGe detector and completely cover passive protection. Active protection lowers integral countdown in the background of a factor of three for the range from 50 to 2800 keV, which lowers the threshold of detection and is suitable for the measurement of environmental samples [17]. For cereal samples the detectors were calibrated with a secondary reference radioactive material in plastic boxes (volume 125 cm^3) obtained from the primary reference radioactive material—Czech Metrological Institute, Praha, 9031-OL-427/12, type ERX, total activity 72.40 kBq on 31.08.2012 (^{241}Am, ^{109}Cd, ^{139}Ce, ^{57}Co, ^{60}Co, ^{203}Hg, ^{88}Y, ^{113}Sn, ^{85}Sr ^{137}Cs, ^{210}Pb) [19].

The counting time was 60,000 s. The results are presented with the expanded measuring uncertainty for the factor k = 2, level of confidence for normal distribution 95%.

2.3 Transfer factor calculations

Transfer factor (TF) was calculated according to Eq. (1), defined as the ratio of specific activity of radionuclide in plant (Bq/kg dry matter) and specific activity in soil (Bq/kg) [18]:

$$TF = \frac{A_p}{A_s} \tag{1}$$

where A_p is the specific activity of the radionuclide in plant [Bq/kg dry matter], and A_s is the specific activity of the radionuclide in soil [Bq/kg].

The change absorbed dose intensity into absorbed dose rate of gamma radiation from the natural radionuclides in soil was calculated according to Eq. (2).
The annual effective dose was calculated according to Eq. (3).

$$D(nGyh - 1) = 0.462 \times CRa + 0.604 \times CTh + 0.0417 \times CK \tag{2}$$

where CRa is the specific activity of ^{226}Ra in soil, CTh is the specific activity of ^{232}Th in soil, and CK is the specific activity of ^{40}K in soil.

$$DE(mS\nu) = 0.7S\nu Gy - 1 \text{ x } 0.2 \text{ x } 365 \text{ x } 24 \text{ x } D \tag{3}$$

3. Results and discussion

The results of the gamma spectrometry analysis of soils at different locations (sampling sites) are presented in **Table 2**.

The results of the calculated absorbed dose intensity and the annual effective doses from natural radionuclides in soils are presented in the table and in **Table 3.**

There are no significant differences among the specific activities of natural radionuclides in soils regarding the sampling depth of the soil at the specific location, i.e., the differences are within the measuring uncertainty. The same applies for the specific activities of ^{137}Cs—their values do not differ significantly regarding the sampling depth of the soil at the specific location. As it has been detected only in traces, it does not present a risk of being accumulated in plants and human diet [20].

For all of the locations, the specific activities of ^{226}Ra are in the range of 22–45 Bq/kg, while for ^{232}Th the values are in the range of 29–55 Bq/kg. For ^{40}K, the specific activities cover the interval from 460 to 730 Bq/kg, for ^{238}U the activities are in the range of 22–51 Bq/kg, and for ^{235}U in the range of 1.1–2.7 Bq/kg. The specific activities of ^{137}Cs cover the interval of 7.2–17 Bq/kg. The uneven distribution of cesium within the same area is mainly due to the relocation and washing out effects in the soil.

There are no significant differences among the specific activities of natural radionuclides between the locations (**Table 2**). The lowest values of the specific activities for ^{226}Ra, ^{232}Th, ^{238}U, and ^{235}U are obtained at Bujkovac, the highest ones at Korbevac. The values are within the range of the literature data of the specific

	Bq/kg					
Depth (cm)	^{226}Ra	^{232}Th	^{40}K	^{238}U	^{235}U	^{137}Cs
			Korbevac			
0–5	43 ± 3	55 ± 4	730 ± 50	47 ± 8	2.7 ± 0.2	16 ± 1
0–10	45 ± 3	54 ± 4	730 ± 50	51 ± 9	2.4 ± 0.2	16 ± 1
0–20	38 ± 3	51 ± 4	690 ± 40	40 ± 8	2.4 ± 0.2	15 ± 1
			Suvi Dol			
0–5	38 ± 3	52 ± 4	490 ± 30	35 ± 8	1.7 ± 0.1	10.1 ± 0.7
5–10	33 ± 2	48 ± 3	470 ± 30	34 ± 9	1.7 ± 0.2	7.9 ± 0.6
10–15	37 ± 3	50 ± 3	460 ± 30	34 ± 8	1.9 ± 0.2	7.2 ± 0.5
			Bujkovac			
0–5	22 ± 2	30 ± 2	500 ± 30	25 ± 8	1.6 ± 0.2	17 ± 1
0–10	23 ± 2	30 ± 2	510 ± 30	25 ± 7	1.5 ± 0.1	18 ± 1
0–20	25 ± 2	29 ± 2	520 ± 30	22 ± 8	1.1 ± 0.1	17 ± 1

Table 2.
Specific activity of radionuclides in soil samples at different depths and sampling sites [Bq/kg].

Depth (cm)	D(nGyh^{-1})	D$_E$(mS$_v$)
Korbevac		
0–5	83.53	0.102
0–10	85.85	0.105
0–20	73.89	0.091
Suvi Dol		
0–5	69.39	0.085
5–10	63.84	0.078
10–15	66.48	0.081
Bujkovac		
0–5	49.13	0.061
0–10	50.75	0.062
0–20	50.01	0.061

Table 3.
Absorbed dose intensity D(nGyh^{-1}) and the annual effective doses D$_E$(mS$_v$) from natural radionuclides in soils.

activities of natural radionuclides in soils reported for the region of former Yugoslavia [6]. Compared to the other locations, the specific activity of ^{226}Ra is lower only in soils sampled at Bujkovac.

The values of the calculated absorbed dose intensity are in the range of 49.13–85.85 nGy/h, while the annual effective doses range from 0.061 to 0.105 mSv/h and are within the values reported for other regions in the country [6].

The results of the levels of natural radionuclides and ^{137}Cs in cereals are presented in **Table 4**.

Table 5 presents the means of the specific activities of the radionuclides in cereals sampled on the investigated locations.

The specific activity of ^{232}Th (2.6 Bq/kg dry matter) presented in **Table 5** refers only to the sample of wheat from the village of Korbevac. In all the other samples of cereals, the specific activity of this radionuclide is under MDA. The specific activities of ^{238}U, ^{235}U, and ^{137}Cs in all investigated samples of cereals are under the MDA, too.

The values of calculated transfer factors are presented in **Table 6**. The values of the specific activity in soil used to calculate the transfer factors were the mean specific activity of the radionuclides for the different sampling depth at the location.

As some of the obtained values of the radionuclides, specific activities in cereals were under MDA; transfer factors were calculated only for ^{40}K, ^{226}Ra, and ^{232}Th. The calculated values of the transfer factors for cereals indicate that ^{40}K and ^{226}Ra are the main radionuclides transferred into the cereal grain. The TF for ^{40}K (0.144–0.392) are higher than the TF for ^{226}Ra and ^{232}Th by an order of magnitude (0.00–80.074 for TF (^{226}Ra)). The TF for ^{40}K can be rather high, as is known and reported in the literature [4]. Other radionuclides do not accumulate

			(Bq/kg)			
Sample	**^{226}Ra**	**^{232}Th**	**^{40}K**	**^{238}U**	**^{235}U**	**^{137}Cs**
Korbevac						
Wheat	2.2 ± 0.4	2.6 ± 0.8	150 ± 10	< 2	< 0.2	< 0.06
Corn	0.4 ± 0.1	< 0.2	108 ± 7	< 1	< 0.06	< 0.03
Suvi Dol						
Wheat	0.30 ± 0.07	< 0.1	106 ± 7	< 0.6	< 0.04	< 0.02
Corn	< 0.2	< 0.2	68 ± 5	<1	< 0.09	< 0.03
Bujkovac						
Wheat	0.37 ± 0.07	< 0.2	102 ± 7	< 1	< 0.04	< 0.02
Corn	1.4 ± 0.3	< 0.4	89 ± 7	< 2	< 0.1	< 0.06
Barley	1.7 ± 0.2	< 0.2	200 ± 10	< 2	< 0.1	< 0.07

Table 4.
Specific activity of radionuclides in grain samples [Bq/kg dry matter].

Radionuclide	Cereals (Bq/kg)		
	Mean value	**Interval**	
		Min	Max
^{226}Ra	1.06	MDA*	2.2
^{232}Th	2.6	MDA	2.6
^{40}K	118	68	200
^{238}U	The values are under MDA		
^{235}U			
^{137}Cs			

**MDA—minimal detection limit.*

Table 5.
Mean values of the radionuclides' specific activities in cereals [Bq/kg dry matter].

	Transfer factor		
Cereal	$^{226}Ra_p/^{226}Ra_s$	$^{232}Th_p/^{232}Th_s$	$^{40}K_p/^{40}K_s$
		Korbevac	
Wheat	0.052	0.051	0.209
Corn	0.009	—	0.151
		Suvi Dol	
Wheat	0.008		0.224
Corn	—	—	0.144
		Bujkovac	
Wheat	0.016	—	0.200
Corn	0.061	—	0.174
Barley	0.074	—	0.392

Table 6.
Value of transfer factor for cereals.

in the plant in more significant amounts [12]. This is mostly due to the discrimination in uptake of essential and nonessential elements, exhibited by the plant [12]. Also, it is reported that small percentage of the total activity found in the plant is accumulated in the root system, while 1–16% is accumulated in the grain [15]. The addition of phosphate to soil reduces the availability of thorium for root uptake through the formation of phosphate salts that have low solubility [15]. Regression analysis, reported in [15], showed that thorium availability to wheat was negatively related to soil pH and positively related to soil organic matter, cationic exchange capacity, and clay content. In comparison to the literature, it can be seen that the obtained TF for cereals in Pcinja region are in agreement with the results obtained in other parts of the world [7, 12, 16], while they are lower by the order of magnitude in comparison to the TF reported for the plants that are principally grass pasture, where the stem and leaves were analyzed (TF(Ra) = 0.17, TF(Th) = 0.058, TF(K) = 1.3 [17]).

4. Conclusion and recommendation

The specific activities of the radionuclides in soil, at all the investigated locations, were in the range from 22 to 45 Bq/kg for ^{226}Ra, from 29 to 55 Bq/kg for ^{232}Th, 460 to 730 Bq/kg for ^{40}K, from 22 to 51 Bq/kg for ^{238}U, from 1.1 to 2.7 Bq/kg for ^{235}U, and from 7.2 to 17 Bq/kg for ^{137}Cs. The obtained specific activities for ^{236}Ra, ^{232}Th, and ^{40}K in "gajnjaca" soil are in good agreement with the values obtained for the other types of soils [4]. The differences between the specific activities of a radionuclide in soil samples from different depths are within the measuring uncertainties, and the ratio of specific activities for ^{235}U/^{238}U suggests the natural origin of uranium. The activities of radionuclides in cereals also do not differ from the values obtained by other authors. Distribution of radionuclides from the soil into the plant depends on the bioavailability of minerals in the soil, the root structure of the investigated plant and the processes in the plant tissue. The calculated values of TF for cereals indicate that ^{40}K and ^{226}Ra are the main radionuclides that are transferred in cereals. This evaluation is most important for the production of foodstuffs diet with low contents of radionuclides.

On the location of Korbevac, Suvi Dol, and Bujkovac the calculated values of TF for ^{40}K were in the range of 0.144–0.392; for ^{226}Ra the values of transfer factors were in the range of 0.008–0.074. It should be noted that the evaluated activities refer to the content of radionuclides in dry plant matter and that the activities in the fresh plants are on the average four to five times lower due to the water content. For other natural radionuclides and for ^{137}Cs, the TF have not been calculated as the specific activities of these radionuclides in cereals were under the MDA. The results presented in this paper are the preliminary investigations of the contents of radionuclides in soils and cereals in the region of Pcinja. As the transfer factors in the "soil-cereal" system were determined only for the specific type of soil, the investigations should continue for other types of soils and cereals mostly used in animal and human diet. The measurements presented in this manuscript are the first to be conducted in the region of Pcinja, thus providing the results that can be used as a baseline for the future measurements and monitoring.

Author details

Jelena Markovic[1]* and Svetlana Stevovic[2]

1 Environmental Protection, Colledge of Applied Professional, Vranje, Serbia

2 Innovation Center, Faculty of Mechanical Engineering in Belgrade, Serbia

*Address all correspondence to: gogaijeka94@gmail.com

IntechOpen

© 2019 The Author(s). Licensee IntechOpen. This chapter is distributed under the terms of the Creative Commons Attribution License (http://creativecommons.org/licenses/by/3.0), which permits unrestricted use, distribution, and reproduction in any medium, provided the original work is properly cited.

References

[1] Gilmor GR. Practical Gamma-Ray Spectrometry. New York, USA: John Wiley and Sons; 2008

[2] UNCSEAR. (United Nations Scientific Committee on the Effects of Atomic Radiation), Report to the general assembly with Scientific Annexes. Annex B: Exposure of the Public and Workers from Various Sources of Radiation. New York, USA; 2010

[3] Bossew P, Kirchner G. Modeling the vertical distribution of radionuclides in soil. Part 1. The converse dispersion equation revisited. Journal of Environmental Radioactivity. 2004;**73**:127-150

[4] Solecki J, Chibowski S. Determination of transfer factors for ^{137}Cs and ^{90}Sr isotopes in soil–plant system. Journal of Radioanalytical and Nuclear Chemistry. 2002;**252**(1):89-93

[5] Karacan F. The simple radiochemical determination of ^{90}Sr in environmental solid samples by solvent extraction. Journal of Radioanalytical and Nuclear Chemistry. 2011;**288**:685-691

[6] Bikit I et al. Monitoring of Radioactivity in Soil on the Territory of Novi Sad during the Year 2012. Faculty of Science, University of Novi Sad; 2012. (in Serbian)

[7] Selvasekarapandian S et al. Natural radionuclide distribution in soil Gudlaore, India. Applied Radiation and Isotopes. 2000;**52**(2):299-306

[8] Pulhani VA et al. Uptake and distribution of natural radioactivity in wheat plants from soil. Journal of Environmental Radioactivity. 2005;**79**(3):331-346

[9] Schimmack W et al. Spatial variability of fallout– ^{90}Sr in soil and vegetation of an alpine pasture. Journal of Environmental Radioactivity. 2003;**65**(3):281-296

[10] Abu-Khadra et al. Transfer Factor of Radioactive Cs and Sr from Egyptian Soils to Roots and Leafs of Wheat Plant. IX Radiation & Protection Conference, Egypt; 2008

[11] Bikit M et al. Transportation of Natural Radionuclides from the Soil into Plants. Proceedings of XIII International Symposium on Protection against Radiation, Becici. 1995. pp. 221-224 (in Serbian)

[12] Sarap N, Rajacic M, Jankovic M, Nikolic J, Dolijanovic Z, Todorovic D, Pantelic G. The Distribution of the Total Beta and Specific Activities of ^{40}K in Vegetative and Generative Organs of Maize. Proceedings of the XXVIII DZZSCG, Institute of Nuclear Sciences Vinca; 2015. pp. 70-73

[13] Popovic D et al. Contents of radionuclides in soils in Serbia: Dose calculations and environmental risk assessment. Advances in Environmental Research. 2012;**6**:91-134 (in Serbian)

[14] Dragovic S et al. Assessment of gamma dose rates from terrestrial exposure in Serbia and Montenegro. Radiation Protection Dosimetry. 2006;**121**:297-302 (in Serbian)

[15] Roesseler CE, Smith ZA, Bolch WE, Prince RJ. Uranium and radium 226 in Florida phosphate materials. Health Physics. 1979;**37**:269-277

[16] Kim KH, Burnet WC. ^{226}Ra in Phosphate nodules from the Peru/Chile seafloor. Geochimica et Cosmochimica Acta. 1985;**49**:1073-1081

[17] Bikit I, Slivka J, Mrdja D, Zikic-Todorovic N, Curcic S, Varga E, et al.

Simple Method for Depleted Uranium Determination. Japanese Journal of Applied Physics (JJAP). 2003:5269-5273. (in Serbian)

[18] Vandenhove H et al. Proposal for new best estimates of soil-to-plant transfer factor of U, Th, Ra, Pb and Po. Journal of Environmental Radioactivity. 2009;**100**(9):721-732

[19] Leo WR. Techniques for Nuclear and Particle Physics Experiments. Berlin Heidelberg, New York: Springer-Verlag; 1994

[20] Vukasinovic I et al. Depth distribution of ^{137}Cs in anthrosol from the experimental field "Radmilovac" near Belgrade, Serbia. Archives of Industrial Hygiene and Toxicology. 2013;**64**:425-430. (in Serbian)

[21] Džoljić J, Stevović S, Todorović D, Polavder S, Rajačić M, Nikolić JK. Natural and artificial radioactivity in some protected areas of South East Europe. Nuclear Technology & Radiation Protection Journal;**XXXII** (4 (December 2017)). ISSN 1451-3994

Section 2

Metals in Soil - Remediation

Chapter 5

Metal-Contaminated Soil Remediation: Phytoremediation, Chemical Leaching and Electrochemical Remediation

Binessi Edouard Ifon, Alexis Crépin Finagnon Togbé, Lyde Arsène Sewedo Tometin, Fidèle Suanon and Arouna Yessoufou

Abstract

Soil contamination has led to serious land tenure problems, reduction in land usability for agricultural production; as a consequence, food insecurity is nowadays a global challenge. Indeed, with rapid population growth across the world, the food demand for consumption has drastically increased and traditional ways of producing food cannot meet with the actual demand. Industrialization has been acknowledged as a way out to sustain humanity with food. Unfortunately, the later has further turn into a threat to the environment. In effect, several potentially toxic elements (PTE) are being released in the environment and soil systems; and arable or agricultural lands are getting restraint, limited and scarce. Nowadays, there is a consensus on remediating contaminated lands with PTE, mainly inorganic contaminants, metals. The state at which a metal is found in the soil greatly influences its bioavailability, interaction with plants and the level at which it will threaten (toxicity) the environment and thus human. It even defines the remediation approaches to be applied for the soil restoration. This chapter will provide an insight on the occurrence of PTE in the soil, bioavailability and remediation approaches namely phytoremediation, chemical leaching and electrochemical remediation; and finally highlight the future research direction on this topic.

Keywords: metals, bioavailability, soil, contamination, decontamination

1. Introduction

Soil is a balanced and complex system, where plants and microorganisms live and co-operate, thus ensuring, crops and food necessary to sustain life [1]. Natural erosion and human activities are enemies of the soil ecosystem. It has been reported that 25% of the global soils are highly degraded and 44% are significantly degraded [2]. Inorganic and organic pollutants are enemies of soils responsible of its contamination. The contamination of soil by a mixture of

IntechOpen

organic and non-organic pollutants due to various anthropogenic and natural causes is one of the most important issues in soil pollution [3]. It threatens humans and the ecosystem via: direct inhalation or through contaminated soil, food chain, or consumption of contaminated surface and ground water, reduction agricultural land (arable land) and in the food's quality; otherwise, there occur an issue related to the reduction of the marketability of farm products as result of safety concern (phytotoxicity) [4].

Among several pollutants threatening soil are: metals [5, 6], through emissions from the rapidly expanding industrial areas, mine tailings, disposal of high metal wastes like e-wastes, leaded gasoline and paints, land application of fertilizers, animal manures, sewage sludge, pesticides, wastewater irrigation [7, 8]; and metalloids [9–11] from industrial waste [12] or mine ores [13]. To be noticed, there are also organic contaminants among which persistent organic pollutants (POP) such as chlorinated [14] and polycyclic aromatic compounds (PAHs) [15], pesticides and herbicides [16] that threaten soil and environment system. Particularly, potentially-toxic elements (PTE) in water and soil have been of great environmental concern due to their non-biodegradable nature, toxicity, bioaccumulation in the food chain, persistence in the environment, and adverse effects on organisms and humans. Chromium (Cr), copper (Cu), nickel (Ni), mercury (Hg), cadmium (Cd), and lead (Pb) are among the environment most concerned toxic PTE. The presence of toxic metals in soil can severely inhibit even the biodegradation of organic contaminants [17]. The treatment thus, protection and remediation of soil are of paramount importance nowadays.

Overwhelming numbers of soil remediation technologies have been developed and tested in both field and controlled environment experiments. Among many, bioremediation (use of microorganism) [18], phytoremediation (use of plants species) soil washing (use of inorganic and organic acids or organic chelators or surfactants), solidification, stabilization, excavation, and electroremediation techniques [19, 20] approaches are commonly used for the treatment of contaminated soil. However, these approaches seem limited and not efficient and effective under severe contamination such as metallic elements and POPs co-contaminated site (e.g. e-waste disposal site or industrial contaminated sites) as microorganisms and plants growth is severely inhibited [17]. Electrokinetic remediation approach which consists in applying direct low level current between two electrodes is nowadays widely used for soil treatment due to it many advantages. The latter shows promising in the future of soil remediation mainly it combination with other technologies; it being under intense investigation. In this chapter of the book we are going to give an insight on the functioning of each of these three approaches during soil treatment, it advantages and limits; and then the direction to explore for a better future of soil remediation.

2. Main sources of metals in the soil

Soil, originally, acts as both source and reservoir of metallic elements [21]. PTE are naturally occurring throughout the earth's crust. However, when talking of soil contamination, nowadays, it refers to the contamination related to anthropogenic activities which led to the increase of contaminants in the soil system; even beyond the threshold concentrations stated in regulations for the safe use of soil in agricultural productions. As consequence, due to its severe contamination, soil represents a major main through which metals are spread in different environment compartments including groundwater, plants, river etc.

Several sources can contribute to soil contamination by metallic elements. Indeed, with the rapid development and industrialization in many countries around the world, there occur an excessive use of various chemical based pesticides and fertilizers in agricultural fields, which results in to the accumulation of PTE in soil and the emerging of serious soil contamination issue [22]. The application of mineral and organic fertilizers can introduce PTE into the soil–plant system. It is commonly known that phosphate rock fertilizers often contain potentially toxic trace elements including copper (Cu), zinc (Zn), manganese (Mn), lead (Pb), and cadmium (Cd) [23, 24]. Several PTE, such as Cadmium (Cd), Chromium (Cr), Copper (Cu), Mercury (Hg), Lead (Pb), Nickel (Ni), Zinc (Zn), and the metalloid Arsenic (As), are widely used by industries, agriculture and consequently released into the environment [25]. Mining is considered to be one of the most significant sources of PTE [26, 27]. In China, it was reported that 1.5 million ha of waste land was the result of PTE contamination caused by mining. Furthermore, area of polluted land keeps increasing at a rate of 46,700 ha/year [26].

Otherwise, with the rapid industrialization and urbanization, the world is facing growing environmental issues [28] with respect the production and disposal of huge amounts of sewage sludge. Indeed, it is noteworthy that huge amount of sewage sludge is being produced yearly and it management remains challenging. Nowadays, one of the mains for the disposal of this matter, is through land application as soil amendment; because the matter is a rich source of phosphorous and nitrogen, and could be value-added as fertilizer [29]. Unfortunately this matter is generally loaded with various pollutants among which metallic elements at a high concentration; which threatens the safety of the receiving soil [30], with its adverse impacts on human and other living organisms when their bioavailability exceeds the concentration. These metals mainly originate from the aqueous phase of the wastewater, and then concentrate in the sludge during the treatment processes like precipitation, coagulation, adsorption etc. Recently, a studied was conducted in China by [31], and over 50 metallic elements including industrial commonly used PTE, rare earth elements and precious metals; were investigated in sewage sludge from different wastewater treatment plants from different region. Results revealed broad range of concentrations of the elements ranging from >125–53,500 mg kg^{-1} dry sludge (DS) for commonly used industrial metals, 1.22–14.0 mg kg^{-1} DS for precious metals, and 1.12–439.0 mg kg^{-1} DS for rare earth elements. The application of such material to soil as amendment would lead to the accumulation and spreading of metals in the soil; mainly with a long-term soil application. Similar result on the occurrence of broad range of metals in the sewage sludge has been reported by [32] with over 60 metals detected in the sewage sludge from different states in US. Overwhelming numbers of reports can be found in the literature regarding the occurrence of metallic elements in the sewage sludge. For example, in 2006, a survey was carried out in china by [33] during which sludge samples collected from over 107 urban wastewater treatment plants (WWTPs) from 48 different provinces across China. Results revealed broad range concentrations of As, Cd, Cr, Cu, Hg, Ni, Pb, and Zn (20.2, 1.97, 93.1, 218.8, 2.13, 48.7, 72.3, and 1058 mg·kg^{-1}, respectively). Another study carried out by [34] reported the present of Cr, Cu, Ni, Pb and Zn in sewage sludge, with concentrations ranging 293.7, 181.7, 114.8, 40.3, 1453.9 mg kg^{-1} DS, respectively. One of the drastic concentration of PTE in the sludge, is the one reported by [35]. Indeed, the author reported higher concentrations up to 172,300, 237, 2225, and 1700 mg kg^{-1} DS for Cd, Cu, Ni and Zn, respectively in an industrial sludge. In addition, [36, 37] recently reported concentration of 64, 73.1, 604.1, 1102.1, 483.9, and 2060.3 mg kg^{-1} DS for Cd, Co, Cu, Cr, Ni and Zn, respectively, in an urban sewage sludge. As can be seen, sewage sludge represent a great sink of metallic pollutants which deserves peculiar attention; as its land application would lead to a drastic soil contamination and metals spreading. This was in accordance with a reported from [38] with respect the application of sewage sludge as soil fertilizer and the risk of metals spreading. To be noticed, aside sewage sludge, poultry and livestock manures from concentrated feeding operations can also, contain PTE and their application to agricultural land can lead to environmental problems and concerns over crop safety.

3. Metal bioavailability, mobility and transport in the soil

It is very important to highlight the fact that potentially-toxic elements (PTE) are not biodegradable elements and can be teratogenic, mutagenic, endocrine disruptors. This means that a metal can only change state of form in the soil; and depending to its forms, it can be transported from soil to another compartment of the environment, and cause serious adverse effects on the environment and human. The behavior and the transportability of a given metal in soil or from the soil to another environment compartment are strongly linked to the state at which the metal is mainly found in the soil. In another word, metal mobility in the soil in strongly linked to their bioavailability. The bioavailability of a metal in the soil is often determined by proceeding a sequential extraction of the metal using various extracting solution. The commonly used sequential extraction procedure is that of Tessier et al. [39]. It consists in to extracting metals in soil in five different fractions including ion exchanges fraction (F2), Carbonate bound-fraction (F3), organic matter-bound fraction (F4) and iron and manganese-bound fraction (F5), and silicate bound/residual fraction (F6). The method has further been modified by introducing a sixth fraction known as water soluble fraction; which normally should be the first fraction (F1) [40]. To be noticed, there are several sequential extraction protocols with various extracting solvents which can be found in the literature. However, following the chemical sequential extraction, metals in soil are generally been extracted in six different fractions (F1–6); which permit to appreciate the state or forms in which a given metal is found and predominate in the soil. Otherwise, the sequential extraction technic permits to evaluate the bioavailability of a metal and thus its mobility in the soil; and finally forecast it potential hazard and toxicity in the environment.

It is widely accepted that the sum of the first three fractions (F1, F2, F3) represents the minimum amount of labile/ bioavailable a given pollutant in the soil that could be easily be mobilized, spread and contaminate the environment [41]. As Result, it is bioavailable for plants uptake. These three fractions are environmental conditions-sensitive [42]. In addition, in the soil system, reactions that often take place are likely to be anaerobic which would lead to the degradation of organic matter in the soil system. As a consequence, the organic matter-bond metals would be released and be redistributed in the soil. This suggests that during the redistribution, the bioavailable fraction of metals could increase, thus increasing their mobility and the risk of environmental contamination. The higher S is for a given metal, the higher are its bioavailability and mobility. It can thus be easily transported in the soil towards the groundwater or be available for plants uptake or washed by runoff and then be transported towards the natural surface water reservoir. So, it can clearly be seen that the more a metal is bioavailable, the lesser its stability in the soil and the higher its toxicity would be. It thus very important to control the bioavailability and mobility of metals in the soil or at some extent, proceed to soil treatment and metals removal.

Otherwise, the bioavailability of a metal in the soil greatly influence it removal. As a consequence, the bioavailability of the metal greatly affects the efficacy and efficiency of soil treatment or remediation technologies [36]. As matter of fact, it is recommended to first take this factor into account before any choice of the treatment or remediation approach. In the following sections, we are going to introduce three main technologies commonly used for soil remediation. It includes phytoremediation, chemical leaching and electrokinetic remediation.

4. Remediation technologies

4.1 Phytoremediation

Phytoremediation refers to the technologies that use living plants including herbs (e.g. *Thlaspi caerulescens*, *Brassica juncea*, *Helianthus annuus*) and woody (e.g. *Salix spp.*, *Populus spp.*) species, to clean up soil, air, and water contaminated with hazardous contaminants using their ability to either contain, remove, uptake, or render harmless various environmental contaminants like potentially-toxic elements, organic compounds and radioactive compounds in soil or water, thanks to their transport capacity and accumulation of contaminants [42, 43]. The use of plants for in situ treatment of contaminated soils was suggested for first time in the early 1990s [44]. The term phytoremediation was then introduced early in the same year to describe the use of plants for extracting PTE from soils [45]. Phytoremediation can be applied to inorganic as well as organic contaminants. As stated by [46], plants are kind of "chemical factories" that exercise great influence on their environment not only by uptake of substances but also by exudation of many molecules that are produced in primary and secondary metabolism. This lively chemical and physical interaction of plants with their environment are of great utility often use for the remediation of contaminated sites; refers to as phytoremediation.

The successful application of phytoremediation techniques is dependent on many parameters among which, contaminants must be bioavailable and ready to be absorbed by roots. The bioavailability of metals depends from solubility of the metals in soil. Nevertheless, mechanisms and efficiency of the phytoremediation depend not only on the bioavailability of metals but also on several others factors such as the nature of contaminant, soil properties, and plant species [47]. The plants which are generally considered for this purpose are those that exhibit great efficiency in phytoremediation processes. They are commonly named as "hyperaccumulator", macrophytes capable of tolerating and accumulating metals present in the soil $\geq$10 g kg^{-1} (1%) Mn or Zn, $\geq$ 1 g kg^{-1} (0.1%) As, Co, Cr, Cu, Ni, Pb, Sb, Se or Tl, and $\geq$ 0.1 g kg^{-1} (0.01%) Cd of the dry mass of shoots on soils rich in PTE in the aerial organs from soils without suffering phytotoxic damage [48]; while yielding low biomass [49]. The List of hyperaccumulators plant species for phytoextraction and phytostabilization has been already in a previous review by Mahar and his co-workers [50].

Otherwise, the extraction efficiency of the pollutants also depends on the biomass produced by the plant. Indeed, the bigger is the biomass the higher the ability of the plant to uptake big quantity of metals. However, more harvests, time and effort will be required to remove the plants after treatment. This will determine the total cost of the entire operation, including disposal, incineration or composting of biomass [51]. Phytoremediation is a reliable reclaiming treatment, because it does not interfere with the ecosystem, it requires less manpower and therefore cost-effective compared to traditional physicochemical methods. This technic knew some significant advancement in recent years thanks to the use of modern biotechnology such as phytoextraction and phytodegradation [51, 52]. Phytoremediation techniques could be applied for the recovery of the industrial sites heavily contaminated with low to moderate concentration.

4.1.1 Mechanisms of phytoremediation

The removal of inorganic pollutants and even organic using phytoremediation is made possible following diverse mechanisms summarized in the **Figure 1** below.

Phytoextraction: metals are extracted from the soil by the plant and transferred to the plant's shoot and leaves. Plants which are often used in this process are selected based on their ability to accumulate contaminants and produce a high biomass [51, 52].

Phytoimmobilization/Phytostabilization: in this process, pollutants are absorbed and immobilized in the root system and it is reduces their mobility. It has been used for the removal of Pb, As, Cd, Cr, Cu and Zn [70, 71].

Phytovolatilization: pollutants are absorbed at root level and converted in a less toxic forms as a result of metabolic modification and released in atmosphere from the aerial parts of plant. We can thus state that this mechanism only relocate the pollutants from the soil to the air [46]. However, in anyway, the soil has been sanitized.

Phytodegradation: this mechanism is mainly for the sequestration of organic contaminants in the soil. It involves Plant enzymes to degrade organic contaminants [51, 52]. Various enzymes are involve in the mechanism among which: (i) dehalogenase (sequestration of chlorinated compounds); (ii) peroxidase (sequestration of phenolic compounds); (iii) nitroreductase (sequestration of explosives and other nitrate compounds); (iv) nitrilase (sequestration of cyanated aromatic compounds); (v) phosphatase (transformation of organophosphate pesticides) [53, 54]. At this

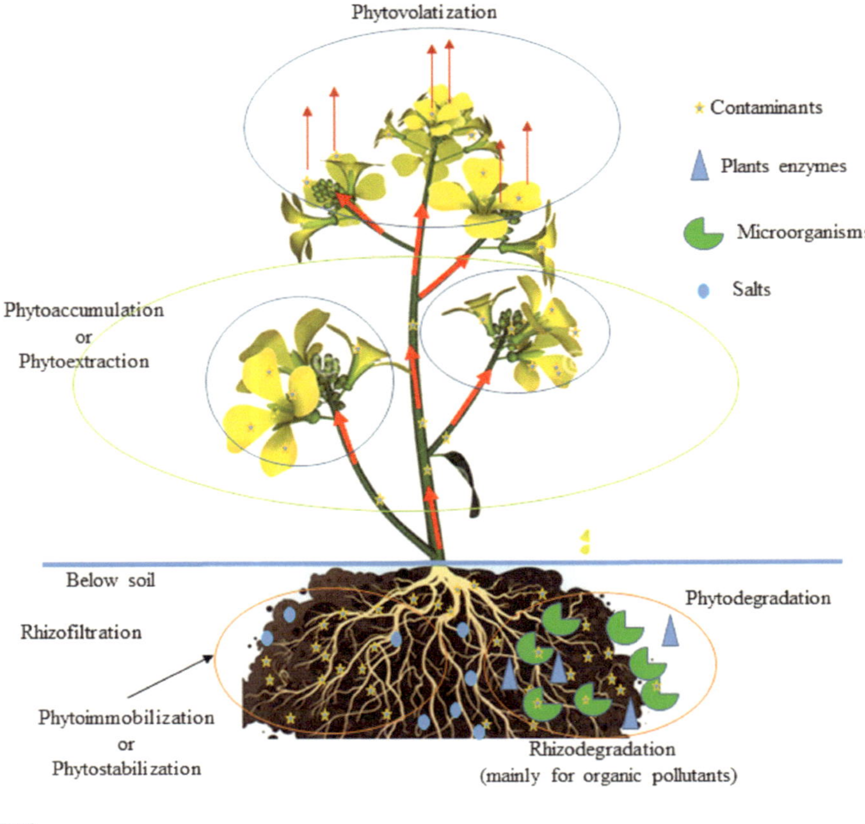

Figure 1.
Different mechanisms involve in phytotechnology.

level, phyto and bioremediation cannot be separated from one another, as microorganisms play an important role in these phytotechnologies. In fact, plants are in continuous interaction with microorganisms, some of which form close associations or symbiotic relationships. This phenomena is what explain the symbiosis that form mycorrhizal fungi with almost all land plants [55] and nitrogen-fixing rhizobia with legumes [56].

Rhizofiltration: this mechanism is commonly applied for the removal of pollutants from surface water or wastewater through adsorption or precipitation on the roots. It has been used for metals and even radioactive elements removal from soil, wastewater and contaminated water with satisfactory results [57]. This technique requires the adjustment of the pH of the medium a better efficiency of the operation; this is seen as a disadvantage of the technique.

Rhizodegradation: just like phytodegradation, this mechanism permit to degradation of organic pollutants in the rhizosphere through rhizospheric microorganisms. It involves a continuous interaction between plants and microorganisms; and thus it cannot be separated from bioremediation. Overwhelming number of research studies has already demonstrated the fact that the number of microorganisms in the rhizosphere is 100 times greater than present on the surface. The latter fetch their nutrients from the root exudates of the plant, which acts as carbon source.

Phytodesalination: this technique is really not used for remediation of contaminated-coil with PTE or persistent organic pollutants but used for the removal of slat from salt-affected soil; it is made possible using halophyte plants (*Artemisia argyi*, *Limonium bicolor*, *Melilotus suaveolens* and *Salsola collina*). Halophytes are plants with great ability to tolerate high concentrations of Na^+ and Cl^- ions; making them able to reclaim excessive saline soil [58]. To be noticed, it is reported that saline soils cover about 6% of the world's land [59] and it well known that salinity is the main environmental factor limiting plant growth and productivity.

4.1.2 Advantage and disadvantage of phytoremediation

In comparison to many other remediation technologies, phytoremediation is found to be of low costs, it protects the soil from erosion (reduction of erosion rate), improves the chemical, physical and biological soil properties, and enhances land esthetic. Phytoremediation is a technology that meets consensus and is highly accepted by the population. It is suitable for sites with low to moderate contamination and where contaminants diffused over large areas, and where there are no temporal limits to the intervention, and finally, it requires less human power. However, despite all this advantages, phytoremediation presents also some limitations which are worth to be mentioned. Indeed, it is time consuming, strong dependence upon: climatic conditions, contaminant(s) concentration and bioavailability, plant tolerance to contaminants, contamination area extent and depth (limited by the rhizosphere or the root zone). The disposable of harvested wastes is another challenge of phytoremediation. It is also not suitable for severely contaminated site such as e-waste contaminated site where potentially-toxic elements and persistent co-exist (the growth of plant would be inhibited), it is also not suitable when arable land (usable land for agricultural production is limited) [60]. Therefore, at this stage, another technology would be need to tackle the remediation of the site. For a better performance of phytoremediation, it could also be combined to electrochemical process. However, the challenge is that the combination would somehow inhibit some phytoremediation processes such as phytodegradation, rhizodegradation which only take place with continuous soil's microorganisms. Indeed, the electrochemical process which includes the induction of low level direct current in the soil via electrodes, would provoke the rising of soil's temperature and the change of soil pH; and thus disturb or inhibit the activity of bacteria. As a consequence, the performance of plant to remove the contaminants will be affected. The detail about electrochemical process, would later be discussed, as it is part of our goal in this chapter.

4.2 Chemical leaching

4.2.1 Chemical leaching and leaching agents

Chemical leaching is one of the traditional remediation technologies used for contaminated soil remediation; and it involves dissolution, extraction and separation of the pollutants. Chemical leaching is one of the common and widely used methods for soil and sludge's PTE removal. Through the precipitation, ions exchange, chelation or adsorption, the PTE in soil are transferred from soil to liquid phase, and then separated from the leachate [61]. The separated pollutants are then converted to the appropriate form before disposal or can be reinserted in the recycling circle. For the dissolution and extraction process, there must be a step of breaking the bound between metals and soil constituents. The success this operation requires the use of acids, oxidants and complexants. Originally, contaminated soil is treated with strong inorganic acids such as HCl, HNO_3, H_2SO_4, H_3PO_4 [62]. Unfortunately, the application of the above-strong acids have been found to be environment and ecological disastrous. Indeed, strong acids have a strong capacity of destroying soil structure, and killing soil's microorganisms. Otherwise, in the process of sanitizing the soil using strong acids, there also occur the loss of soil constituent which is of great concern for the ecological consideration. Such situation is not in line with the protection of the environment on one hand, and does inhibit the productivity of the treated soil on the other hand. As a consequence, the use of strong acids is not environmental friendly. Thus, the integrated utilization of acids or reagents should be deliberately selected to fulfill the requirement of target contaminants removal on one hand, and soil ecological protection on the other hand. This justifies the introduction of Low molecular weight organic acids such as acetic acid, oxalic acid, which constitute a group of weak organic acids [63] and chelating reagents such as nitrilotriacetic acid (NTA) [33], sodium tripolyphosphates (STPP) and ethylenediaminetetraacetic acid (EDTA) [33, 63]. The use of weak acids showed mitigated results even though promising. On the other hand, chelating agents develop great affinity with the metals ions and possess prominent properties of oxidizing and forming complexes with metals cations; which could improve their extraction efficient. The use of the mentioned organic chelators has been widely investigated and results are satisfactory; mainly EDTA is well known for its excellent ability to recover metals from soil (25–80%) depending on the type of soil [64, 65]. However, these chelators seem to be refractory to the environment, and not easily biodegradable and thus can pose a secondary pollution via leaching to the groundwater [66]. As a consequence, there is a need to find more suitable chelators for the replacement of the refractory ones. In line with this objective, the use of organic acids and new generation of chelating agents are increasingly been investigated as an alternatives to above-mentioned washing reagents. N, N-bis(carboxymethyl) glutamic acid (GLDA), a chelator with excellent biodegradability [67], more than 60% degradable within 28 days. According to the OECD 301D test [68] with lowest 'eco-footprint' characteristics in comparison to EDTA and STPP; has been suggested due to it exceptional chelating capacity towards different divalent metal ions [69]. It was successfully used by [35] and [36] for the recovery of Cd Co, Cr, Cu, Ni and Zn from dewatered sewage sludge. The removal efficacy was comprised between 60 and 86% and 70–94% for both studies, respectively. In addition, it comparison with citric acid during the work of [36] showed great efficacy and efficiency of GLDA compared to citric acid. The more a chelators possesses a carboxyl group (-COOH), the higher its performance would be during soil washing process. However, to be noticed, the overwhelming number of research work carried out on this topic which can be found in the literature are lab scale experiments, which is much easier to proceed comparing to field demonstration, mainly *in situ* application. It is only used in an ex-situ remediation technology, which create too much disturbance of soil system and its microorganisms. Here below (**Table 1**) are some organic chelators used in soil washing technology.

4.2.2 Challenges related to field application of chemical leaching

During chemical leaching, the use of significant amount of chelating agent is essential for the mobilization of PTE within the soil system. The addition of chelants to soils not only promote metals mobilization and transfer from the soil to the chelants' solutions but it also increases the total concentration of the soluble metals. A better mobilization of metals in the soil, requires up to hundreds of mill molar per liter concentration of the chelating agents in the soil solution. The issue is that the process can recover only part of the concentration of the dissolved metals, and leaching will be unavoidable [70]; which could lead to the possible contamination of the ground water and slow (several weeks or months) decomposition of the synthetic organic acids. Following the application of chelate forming agents, the removal of metals may continue for a long time. Besides, the use of chelating agents could exercises adverse effects on the soil microorganisms [71].

Otherwise, except the fact that during the soil washing/leaching process, soil minerals and other constituents are washing away together with the target pollutants, the *in situ* application of this technology at the large scale would be very challenging. Indeed, the injection of washing reagent in the soil is really challenging as it would not be easy to control the flow direction; and the solution will tend to flow vertically (leaching towards ground water) rather than in the desired direction, generally horizontal. As a consequence, the *in situ* field applicability of the technology at the large scale is limited; only *ex-situ* application are widely known. Otherwise, the technology is solvent consuming and involve longue processes and post treatments of the treatment waste and thus time consuming with high requirement of human power. Otherwise, it is soil generate too much soil disturbance (soil returning). One of the alternative to make valuable this technology is to combine it with other technology which permit the control of the solvent flow with less soil disturbance such as electrochemical process. This combination has given birth to the electrokinetic remediation technology.

4.3 Electrokinetic remediation

4.3.1 Principles and mechanisms of inorganic contaminants removal in soil

Electrokinetic remediation is a technique that consists in displacing or moving pollutants in contaminated soil from their contaminated points towards a specific controlled extraction points which are generally the electrodes cells. This technique is made possible by the application of a direct low current between electrodes well-disposed in the soil in order to optimize the electric field. The principle of pollutants cleanup is controlled by some key processes such as electroosmosis, electromigration and electrophoresis [72]. These mechanisms involve different mechanism. **Electroosmosis** knows as electroosmotic flow, consists of the

Name	Molecular structure	Name	Molecular structure
Ethylenediaminetetraacetic acid (EDTA) Molecular weight: 292.24 g/mol, appearance: colorless crystal, density: 0.860 g/mL at 20°C, solubility in H_2O: in any ratio biodegradability: moderate	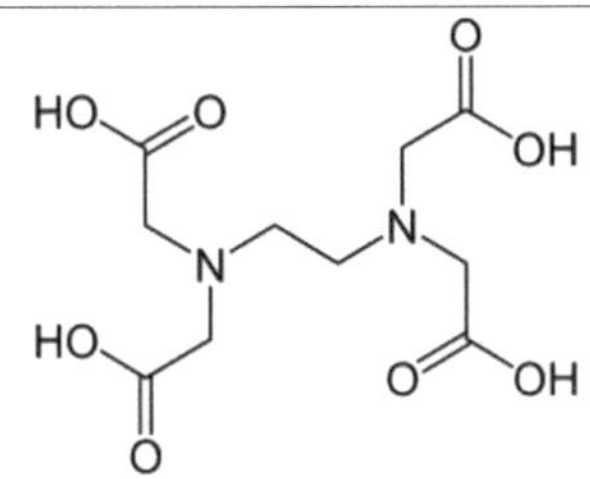	Ethylenediaminedisuccinic acid (EDDS) Molecular weight: 358.1 g/mol, appearance: colorless to yellowish, pH: 9.2 density: 1.26 g/mL, solubility in H_2O: in any ratio biodegradability: > 60%	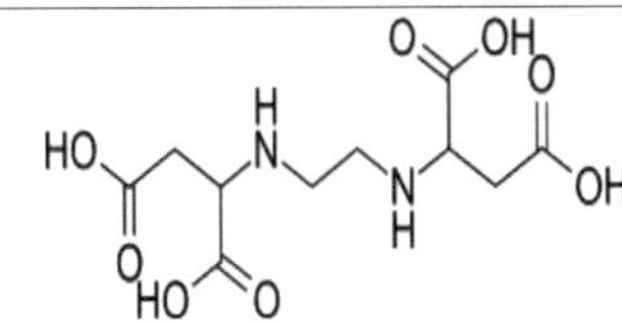
Nitrilotriacetic acid (NTA) Molecular weight: 191.14 g/mol, appearance: white crystal, density: 1.6 g/mL solubility in H_2O: insoluble(<0.01 g/100 mL), biodegradability: easily biodegradable	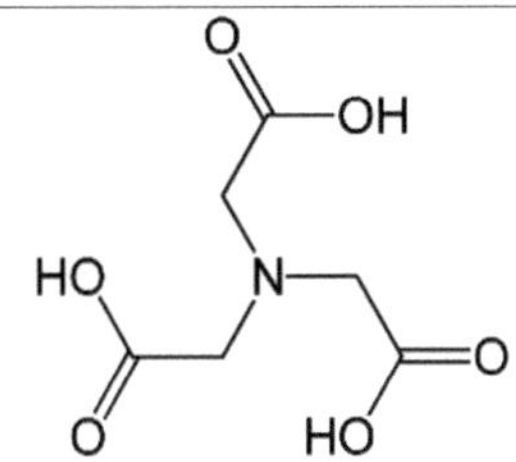	Methylglycinediacetic acid (MGDA) Molecular weight: 271.0 g/mol, appearance: clear yellowish, pH: 11.0 density: 1.31 g/mL, solubility in H_2O: in any ratio biodegradability: > 68%	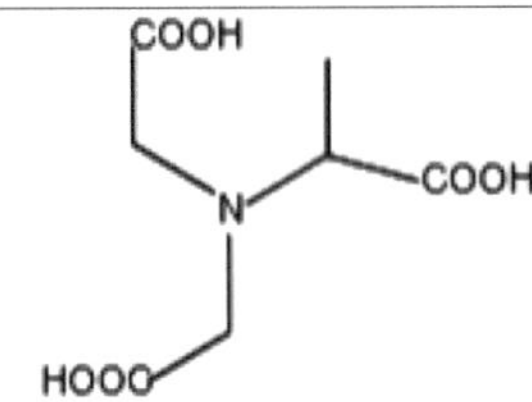
N, N-bis(carboxymethyl) glutamic acid tetra sodium salt (GLDA) Molecular weight: 351.1 g/mol, appearance: colorless to yellowish pH: 13.5, density: 1.38 g/mL, solubility in H_2O: in any ratio biodegradability: > 83%	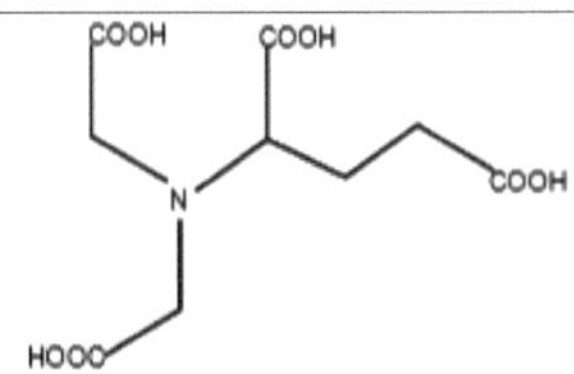	N-(1,2dicarboxyethylene)D,L-asparagine acid (IDS) Molecular weight: 337.1 appearance: colorless to light yellow pH: 10.3–11.4 density: 1.32–1.35 g/mL solubility in H_2O: in any ratio biodegradability: > 80%	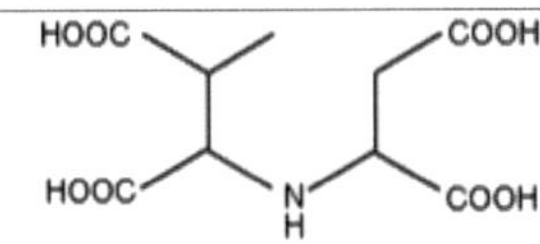

Table 1.
Some organic chelators often used for soil washing, EDTA and NTA are commonly used, while others in the table are known as new generation of chelators [69].

displacement of the liquid in the porous soil as result of the application of the electric field. During this movement, the pore fluid carries along organics and neutral molecules. **Electromigration** consists of the transport of charged particles (anions and cations) towards the opposite electrode cell. As for the **electrophoresis**, it is the movement of dispersed particles in the medium relative to a fluid as result of a spatially uniform electric field. These mechanisms are of great importance in pollution remediation (soil and sediment treatment) when using electrokinetic approach.

During electrokinetic remediation, there occur electrochemical reactions of which, electrolysis of water represents one of the most important and influential reactions. These reactions take place on the surface of the electrodes as the result of the application of low direct electric current. During electrolysis process, there occur a generation of protons ($H+$) on the anodic surface and hydroxyl ions ($OH-$) on the cathodic surface; which lead to an important pH gradient (**Figure 2**). These ionic species are mobilized through the soil at a rate determined mainly by the electromigration and diffusive processes and the soil's buffering capacity [73].

The pH profile is a key parameter during soil treatment with electrokinetic approach. Indeed, the changes of pH induce beside electrokinetic processes, physicochemical processes among which precipitation/dissolution of minerals and metals, adsorption/desorption of pollutants and ion exchange between the soil solid and the pore water. As it is well known, pH exercises strong influence on the chemical speciation of the compounds mainly inorganic present in the soil system. It determines the state or ionic forms in which a compound is found in the soil. This will indirectly condition the predominant transport mechanism by which this compound will move during the treatment.

Especially the change in pH affects the surface charge of soil particles and metal ions mobility. The generated acidic conditions help mobilize sorbed metal ions, prevents formation of metal hydroxide and carbonate precipitates; and thus facilitate their electromigration via the electroosmotic flow of the liquid. However, highly acidic conditions cause electroosmotic flow to stop or reverse, whereas alkaline condition results in PTE precipitation and increases electroosmotic flow.

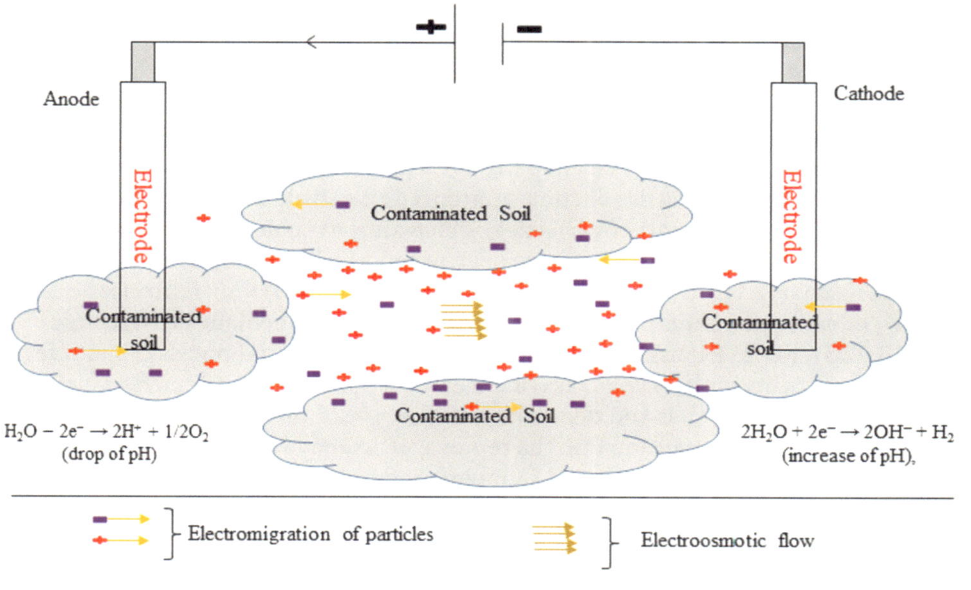

Figure 2.
Mechanism of electrokinetic remediation approach.

Thus, to maintain this parameter within a suitable range, pH control if often performed in both anode and cathode by adding sodium hydroxide (0.1 and 1 M) and acetic acid/citric acid (0.1 and 1 M) respectively [74, 75]. The in-situ acidification, however, may not be adequate if the soil possesses high buffering capacity. Moreover, the generated base front causes metal ions to precipitate, impeding their final arrival at the cathode [76]. Consequently, external/artificial acidification is often required even necessary during electrokinetic soil remediation [77]. However, the use of strong inorganic acids such as HCl, HNO_3 is not is not recommended as it can damage the soil structure. In addition, it would be costly and is not environmentally acceptable. Generally, water or chemical solutions [(0.1 M) EDTA or acetic acid, citric acid, etc.] are continuously injected at the anode to maintain optimal remediation conditions; contaminated water is removed at the cathode by pumping [78].

This technology has been successfully used in single for the treatment of various wastes/sites such as wastewater, sewage sludge, soil and sediments contaminated with inorganic and organic pollutants [76, 77, 79]. However, to optimize its efficacy, it has also been used in the combinations with other technologies [80–82]. The combination of electrokinetic remediation method with other technologies has been tested and is still on the hotspot of scientific research in environmental filed. It includes electrokinetic-microbe joint remediation, electrokinetic-chemical joint remediation [82], electrokinetic-oxidation/reduction joint remediation [83], coupled electrokinetic-phytoremediation [81], electrokinetics coupled with electrospun polyacrylonitrile nanofiber membrane [80], and electrokinetic remediation conjugated with permeable reactive barrier [79].

4.3.2 Electrodes and electrolytes

Various inert electrodes made of ceramic, carbon, graphite, titanium, stainless steel, are generally used during electrokinetic remediation of contaminated-soil. Each electrode has its level of stability, the choice of electrode depends on the use and purpose. The electrode are configured in order to optimize the electrical field in the treated area. Generally, they are disposed in the contaminated soil at 1.0–1.5 m spacing, with imposed DC current at 1.0–3.0 V cm^{-1} or 100–500 kWh m^{-3} [84].

Electrokinetic extraction of PTE involves desorption/dissolution followed by transport. When the concentration of PTE in the soil solution becomes below the soil sorption capacity, chemical additives are typically needed to help mobilize and sorb metals. Also poor conductivity-pollutants (in the form of sulfides) or present in metallic form (Hg) cleanup involve a primary step of dissolution. This step generally involves the use of some appropriate electrolytes such as distilled water, organic acids or synthetic chelates; which aims to enhance the efficiency of the remediation. Several chemical have been tested as additives and include acetic acid (CH_3COOH), citric acid ($(HOOC\text{-}CH_2)_2C(OH)(COOH)$), nitrilotriacetic acid (NTA), ethylene-diamine-tetra-acetic acid (EDTA), ethylenediaminedisuccinic acid (EDDS), diethylenetriaminepentaacetic acid (DTPA), and potassium iodide (KI). These additives also known as enhancement fluids mobilization efficiencies varies from one to another and depending on the type of metal species in soil [85–87]. It is worth to mention that the removal efficiency varies not only depending on the type of the chemical used (anolyte) and metal remediated [88] but also on the type of electrode. Indeed, the use of KH_2PO_4 as an anolyte permitted to enhance the removal efficiencies of As species by >50% and ~ 20% for Cu species. Meanwhile, it did not enhanced the removal of the Pb and Zn (< 20%) [89, 90]. Also reported that adding ethylene diamine disuccinate (EDDS) in the anolyte enhanced Pb and Cd removal efficiencies in the contaminated soil.

4.3.3 Advantages and limitations

Electrokinetic technology has many advantages among which, it applicability for in-situ/ex-situ remediation, applicable to low-permeability soils and a mixture of contaminants where other technologies cannot be applied, applicable to a wide range of pollutants, and applicable to heavy and severely contaminated sites. However, the main limiting factor for direct electrokinetic remediation is the fluctuation in soil pH; because it cannot maintain soil pH value. Therefore there is a need to control the soil pH by external intervention through the addition of buffer solutions in cathode and anode cells. In fact, controlling the pH in the electrode cells remains the main challenge of this technology. Electrokinetic remediation has shown promising results and is still under development stage [91].

5. Comparison of the three technology

The comparison of the three technologies involved in the present chapter is summarized in the **Figure 3** below.

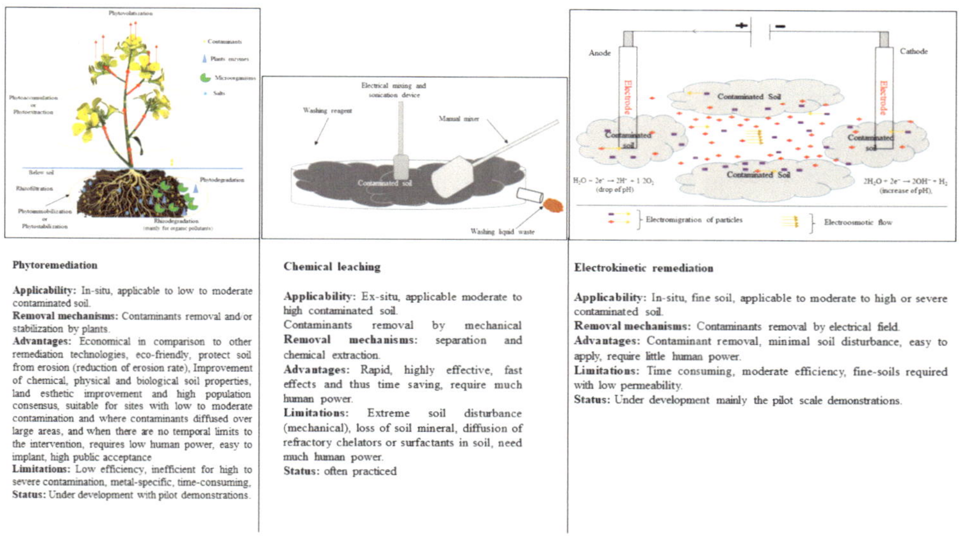

Figure 3.
Comparison of phytoremediation, chemical leaching and electrokinetic technologies.

6. Conclusion

Soil contamination is one of the greatest challenges threatening the world as it lowers soil productivity and compromises food security. Contaminated soil/sites remediation or restoration is among the top list objectives of Food and Agriculture Organization's (FAO) agenda. Phytoremediation, chemical leaching and electrochemical remediation are three techniques commonly used for the remediation of contaminated sites. Each of these techniques has its advantages and limitations. Due to the non-availability of enough arable land, the use of phytoremediation, though it is eco-friendly, would lead to food insecurity as it takes long period to clean a target site. Moreover, it takes too much agricultural space for its implementation. As for chemical leaching, it is an *ex-situ* treatment technique, it thus disturbs too much the soil and its microorganisms; it leads to the loss of much soil minerals and reducing soil fertility (non-suitable for agricultural land). In addition, it introduce much chemical in to the soil, some of which may be refractory to biodegradation and leach to underground water. Electrokinetic approach is less time consuming and less disturbs the treated site; the main challenge is how to control the pH during the process; this could be monitor by external intervention. However, additives which include surfactants, chelants and organic acids must be carefully chosen having in mind their biodegradability and the protection of the soil structure and ecosystem. None of these techniques, when applied in single, is able to properly achieve the soil depollution; thus their combination is highly recommended. The combination of these technologies still suffer some lack of information which need to be explored in order to appreciate their feasibility. In order to enhance the efficiency of soil remediation, it is recommended to investigated and develop more environmental friendly flushing reagents to replace refractory existing ones on one hand; and to promote phyto-electrokinetic remediation approach on the other hand.

Conflict of interest

None.

Author details

Binessi Edouard Ifon[1,4], Alexis Crépin Finagnon Togbé[1], Lyde Arsène Sewedo Tometin[3], Fidèle Suanon[1,2]* and Arouna Yessoufou[4]

1 Laboratory of Physical Chemistry, University of Abomey-Calavi, Cotonou, Republic of Benin

2 State Key Laboratory of Soil Environmental Chemistry and Pollution Remediation, Institute of Soil Science, Chinese Academy of Sciences, Nanjing, PR China

3 Laboratory of Inorganic Chemistry and Environment, University of Abomey-Calavi, Cotonou, Benin

4 Laboratory of Applied Hydrology, University of Abomey-Calavi, Cotonou, Benin

*Address all correspondence to: fidele.suanon@fast.uac.bj

IntechOpen

© 2019 The Author(s). Licensee IntechOpen. This chapter is distributed under the terms of the Creative Commons Attribution License (http://creativecommons.org/licenses/by/3.0), which permits unrestricted use, distribution, and reproduction in any medium, provided the original work is properly cited.

References

[1] Floris B, Galloni P, Sabuzi F, Conte V. Metal systems as tools for soil remediation (review). Inorganica Chimica Acta. 2017;**455**:429-445

[2] The State of the world's Land and Water Resources for Food and Agriculture, Managing Systems at Risk. Rome: FAO; 2011. http://www.fao.org/docrep/017/i1688e/i1688e00.htm [Accessed on 11/6/2018]

[3] Saberi N, Aghababaei M, Ostovar M, Mehrnahad H. Simultaneous removal of polycyclic aromatic hydrocarbon and potentially-toxic elements from an artificial clayey soil by enhanced electrokinetic method. Journal of Environmental Management. 2018;**217**:897-905

[4] Ling W, Shen Q, Gao Y, Gu X, Yang Z. Use of bentonite to control the release of copper from contaminated soils. Australian Journal of Soil Research. 2007;**45**(8):618-623

[5] Vodyanitskii YN, Plekhanova IO. Biogeochemistry of potentially-toxic elements in contaminated excessively moistened soils (analytical review). Eurasian Soil Science. 2014;**47**(3):153-161

[6] Henry H, Naujokas MF, Attanayake C, Basta NT, Cheng Z, Hettiarachchi GM, et al. Bioavailability-based In situ remediation to meet future Lead (Pb) standards in urban soils and gardens. Environmental Science & Technology. 2015;**49**(15):8948-8958

[7] Khan S, Cao Q, Zheng MY, Huang ZY, Zhu GY. Health risks of potentially-toxic elements in contaminated soils and food crops irrigated with wastewater in Beijing, China. Environmental Pollution. 2008;**152**(3):686-692

[8] Zhang KM, Liu YZ, Wang H. Use of single extraction methods to predict bioavailability of potentially-toxic elements in polluted soils to rice. Communications in Soil Science and Plant Analysis. 2010;**41**:820-831

[9] Bolan N, Kunhikrishnan A, Thangarajan R, Kumpiene J, Park J, Makino T, et al. Remediation of potentially-toxic elements(loid)s contaminated soils to mobilize or to immobilize? Journal of Hazardous Materials. 2014;**266**:141-166

[10] Hettick BE, Cañas-Carrell JE, French AD, Klein DM. Arsenic: A review of the Element's toxicity, plant interactions, and potential methods of remediation. Journal of Agricultural and Food Chemistry. 2015;**63**(32):7097-7107

[11] Pierart A, Shahid M, Séjalon-Delmas N, Dumat C. Antimony bioavailability: Knowledge and research perspectives for sustainable agricultures. Journal of Hazardous Materials. 2015;**289**:219-234

[12] Kříbek B, Majer V, Knésl I, Keder J, Mapani B, Kamona F, et al. Contamination of soil and grass in the Tsumeb smelter area, Namibia: Modeling of contaminants dispersion and ground geochemical verification. Applied Geochemistry. 2016;**64**:75-91

[13] Pardo T, Clemente R, Alvarenga P, Bernal MP. Efficiency of soil organic and inorganic amendments on the remediation of a contaminated mine soil: II. Biological and ecotoxicological evaluation. Chemosphere. 2014;**107**:101-108

[14] Achari G, Jakher A, Gupta C, Dhol A, Langford CH. Practical method to extract and dechlorinate PCBs in soils. The Practice Periodical of Hazardous, Toxic, and Radioactive Waste Management. 2010;**14**:98-103

[15] Gan S, Lau EV, Ng HK. Remediation of soils contaminated with polycyclic aromatic hydrocarbons (PAHs). Journal of Hazardous Materials. 2009;**172**(2-3):532-549

[16] Pateiro-Moure M, Arias-Estévez M, Simal-Gándara J. Critical review on the environmental fate of quaternary ammonium herbicides in soils devoted to vineyards. Environmental Science & Technology. 2013;**47**(10):4984-4998

[17] Maslin P, Maier MR. Rhamnolipid-enhanced mineralization of phenanthrene in organic-metal co-contaminated soils. Bioremediation Journal. 2000;**4**(4):295-308

[18] Megharaj M, Ramakrishnan B, Venkateswarlu K, Sethunathan N, Naidu R. Bioremediation approaches for organic pollutants: A critical perspective. Environment International. 2011;**37**(8):1362-1375

[19] Kumpiene J, Lagerkvist A, Maurice C. Stabilisation of As, Cr, Cu, Pb and Znin soil using amendments – A review. Waste Management. 2008;**28**:215-255

[20] Mulligan CN, Yong RN, Gibbs BF. Remediation technologies for metal contaminated soils and groundwater: An evaluation. Engineering Geology. 2001;**60**:193-207

[21] Khan S, Rehman S, Khan AZ, Khan MA, Shah MT. Soil and vegetables enrichment with potentially-toxic elements from geological sources in Gilgit, northern Pakistan. Ecotoxicology and Environmental Safety. 2010;**73**(7):1820-1827

[22] Wuana RA, Okieimen FE. Potentially-toxic elements in contaminated soils: A review of sources, chemistry, risks and best available strategies for remediation. ISRN Ecology. 2011;**2011**:402-647

[23] Dissanayake CB, Chandrajith R. Phosphate mineral fertilizers, trace metals and human health. Journal of the National Science Foundation of Sri Lanka. 2009;**37**:153-165

[24] Zaccone C, Di CR, Rotunno T, Quinto M. Soil-farming system-food-health: Effect of conventional and organic fertilizers on potentially-toxic elements (Cd, Cr, Cu, Ni, Pb, Zn) content in semolina samples. Soil and Tillage Research. 2010;**107**:97-105

[25] Tchounwou PB, Yedjou CG, Patlolla AK, Sutton DJ. Heavy Metal Toxicity and the Environment. In: Luch A, editor. Molecular, Clinical and Environmental Toxicology. Experientia Supplementum, 2012;**101**:133-164. DOI: 10.1007/978-3-7643-8340-4_6

[26] Zhuang P, McBride MB, Xia H, Li N, Li Z. Health risk from potentially-toxic elements via consumption of food crops in the vicinity of Dabaoshan mine, South China. Science of the Total Environment. 2009;**407**:1551-1561

[27] Li P, Lin C, Cheng H, Duan X, Lei K. Contamination and health risks of soil potentially-toxic elements around a lead/zinc smelter in southwestern China. Ecotoxicology and Environmental Safety. 2015;**113**: 391-399

[28] Jin L, Zhang G, Tian H. Current state of sewage treatment in China. Water Research. 2014;**66**:85-98

[29] Roca-Pérez L, Martínez C, Marcilla P, Boluda R. Composting rice straw with sewage sludge and compost effects on the soilplant system. Chemosphere. 2009;**75**:781-787

[30] Nam D-H, Lee B-C, Eom I-C, Kim P, Yeo M-K. Uptake and bioaccumulation of titanium- and silver-nanoparticles in aquatic ecosystems. Molecular & Cellular Toxicology. 2014;**10**:9-17

[31] Suanon F, Sun Q, Yang X, Chi Q, Mulla IS, Mama D, et al. Assessment of the occurrence, spatiotemporal variations and geoaccumulation of fifty-two inorganic elements in sewage sludge: A sludge management revisit. Science Reporter. 2017;7:5698

[32] Westerhoff P et al. Characterization, recovery opportunities, and valuation of metals in municipal Sludges from US wastewater treatment plants Nationwide. Environmental Science & Technology. 2015;**49**:9479-9488

[33] Yang J et al. Current status and developing trends of the contents of potentially-toxic elements in sewage sludges in China. Environmental Science and Engineering. 2014;**8**:719-728

[34] Dong B, Liu X, Dai L, Dai X. Changes of potentially-toxic elements speciation during high-solid anaerobic digestion of sewage sludge. Bioresource Technology. 2013;**131**:152-158

[35] Wu Q, Cui Y, Li Q, Sun J. Effective removal of potentially-toxic elements from industrial sludge with the aid of a biodegradable chelating ligand GLDA. Journal of Hazardous Materials. 2015;**283**:748-754

[36] Suanon F, Sun Q, Dimon B, Mama D, Yu C-P. Potentially-toxic elements removal from sludge with organic chelators: Comparative study of N, N-bis(carboxymethyl) glutamic acid and citric acid. Journal of Environmental Management. 2016a;**166**:341-347

[37] Suanon F et al. Effect of nanoscale zero-valent iron and magnetite (Fe3O4) on the fate of metals during anaerobic digestion of sludge. Water Research. 2016b;**88**:897-903

[38] Suanon F et al. Utilization of sewage sludge in agricultural soil as fertilizer in the Republic of Benin (West Africa): What are the risks of potentially-toxic elements contamination and spreading? American Journal of Environmental Sciences. 2016c;**12**(1):8-15

[39] Tessier A, Campbell PG, Bisson M. Sequential extraction procedure for the speciation of particulate traces metals. Analytical Chemistry. 1979;**51**:844-858

[40] Aikpokpodion PE, Lajide L, Aiyesanmi AF. Characterization of potentially-toxic elements fractions in agricultural soils using sequential extraction technique. World Journal of Agricultural Sciences. 2013;**9**:45-52

[41] Achiba WB, Lakhdar A, Gabteni N, Laing GD, Verloo M, Boeckx P, et al. Accumulation and fractionation of trace metals in a Tunisian calcareous soil amended with farm yard manure and municipal solid waste compost. Journal of Hazardous Materials. 2010;**176**:99-108

[42] Labanowski J, Monna F, Bermond A, Cambier P, Fernandez C, Lamy I, et al. Kinetic extractions to assess mobilization of Zn, Pb, Cu, and Cd in a metal-contaminated soil: EDTA vs citrate. Environmental Pollution. 2008;**152**:693-701

[43] Das PK. Phytoremediation and Nanoremediation: Emerging techniques for treatment of acid mine drainage water. Defence Life Science Journal (DESIDOC). 2018;**3**(2):190-196

[44] Cunningham SD, Berti WR. Remediation of contaminated soils with green plants: An overview. In Vitro Cellular & Developmental Biology. 1993;**29**:207-212

[45] Medina VF, McCutcheon SC. Phytoremediation: An ecological solution to organic chemical contamination. Ecological Engineering. 2002;**18**:647-658

[46] Martin JK. Effect of soil moisture on the release of organic carbon from wheat roots. Soil Biology and Biochemistry. 1977;**9**:1-7

[47] Sreelal G, Jayanthi R. Review on phytoremediation technology for removal of soil contaminant Indian. Journal of Scientific Research. 2017;**14**(1):127-130

[48] Verbruggen N, Hermans C, Schat H. Molecular mechanisms of metal hyperaccumulation in plants. The New Phytologist. 2009;**181**:759-776

[49] Van Oosten MJ, Maggio A. Functional biology of halophytes in the phytoremediation of potentially-toxic elements contaminated soils. Environmental and Experimental Botany. 2014;**111**:135-146

[50] Mahar A, Wang P, Ali A, Awasthi MK, Lahori AH, Wang Q, et al. Challenges and opportunities in the phytoremediation of potentially-toxic elements contaminated soils: A review. Ecotoxicology and Environmental Safety. 2016;**126**:111-121

[51] Sharma P, Pandey S. Status of phytoremediation in world scenario. International Journal of Environmental Bioremediation & Biodegradation. 2014;**2**(4):178-191

[52] Souza EC, Vessoni-Penna TC, de Souza Oliveira RP. Biosurfactant-enhanced hydrocarbon bioremediation: An overview. International Biodeterioration and Biodegradation. 2014;**89**:88-94

[53] Winquist E, Björklöf K, Schultz E, Räsänen M, Salonen K, Anasonye F, et al. Bioremediation of PAH-contaminated soil with fungi - from laboratory to field scale. International Biodeterioration and Biodegradation. 2013;**86**:238-247

[54] Deng Z, Cao L. Fungal endophytes and their interactions with plants in phytoremediation: A review. Chemosphere. 2017;**168**:1100-1106

[55] Harrison MJ. Molecular and cellular aspects of the arbuscular mycorrhizal symbiosis. Annual Review of Plant Physiology and Plant Molecular Biology. 1999;**50**:361-389

[56] Medina VF, McCutcheon SC. Phytoremediation: An ecological solution to organic chemical contamination. Ecological Engineering. 2002;**18**:647-658

[57] Prell J, Poole P. Metabolic changes of rhizobia in legume nodules. Trends in Microbiology. 2006;**14**(4):161-168

[58] Prasad MNV. Sunflower (*Helinathus annuus* L.) - A potential crop for environmental industry. Helia. 2007;**30**(46):167-174

[59] Flowers T, Colmer TD. Plant salt tolerance: Adaptations in halophytes. Annals of Botany. 2015;**115**:327-331. Available online at https://academic.oup.com/aob/article/115/3/327/306278

[60] Bini C. From soil contamination to land restoration. In: Steimberg RV, editor. Contaminated Soils: Environmental Impact, Disposal and Treatment. New York: Nova Science Publishers, Inc; 2009. pp. 97-137

[61] Ferraro A, van Hullebusch ED, Huguenot D, Fabbricino M, Esposito G. Application of an electrochemical treatment for EDDS soilwashing solution regeneration and reuse in a multi-step soil washing process: Case of a Cu contaminated soil. Journal of Environmental Management. 2015;**163**:62-69

[62] Stylianou MA, Kollia D, Haralambous KJ, Inglezakis VJ, Moustakas KG, Maria DL. Effect of acid treatment on the removal of potentially-toxic elements from sewage sludge. Desalination. 2007;**215**:73-81

[63] Zaleckas E, Paulauskas V, Sendžikienė E. Fractionation of potentially-toxic elements in sewage sludge and their removal using low-molecular-weight organic acids. Journal of Environmental Engineering and Landscape Management. 2013;**21**:189-198

[64] Reddy KR, Al-Hamdan AZ, Ala P. Enhanced soil flushing for simultaneous removal of PAHs and potentially-toxic elements from industrial contaminated soil. Journal of Hazardous, Toxic, and Radioactive Waste. 2011;**15**:166-174

[65] Jiang W, Tao T, Liao Z. Removal of potentially-toxic elements from contaminated soil with chelating agents. Open Journal of Soil Science. 2011;**1**:70-76

[66] Nowack B. Environmental chemistry of aminopolycarboxylate chelating agents. Environmental Science & Technology. 2002;**36**:4009-4016

[67] Seetz J, Stanitzek T. GLDA: The New Green Chelating Agent for Detergents and Cosmetics SEPAWA Congress and European Detergents Conference Proceedings. 15-17 October 2008; 2008. pp. 17-22

[68] Schneider J, Potthoff-Karl B, Baur R, Oftring A, Greindl T. Use of glycine-N,N-diacetic acid derivatives as biodegradable complexing agents for alkaline earth metal ions and potentially-toxic elements ions. US Patent 6, 1999; 008-176

[69] Kołodynska D, Robert Y. In: Ning, editor. Chelating Agents of a New Generation as an Alternative to Conventional Chelators for Potentially-Toxic Elements Ions Removal from Different Waste Waters, Expanding Issues in Desalination. 978-953-307-624-9. Rijeka: InTech; 2011

[70] Nowack E, Schulin R, Robinson BH. Critical assessment of chelant-enhanced metal phytoextraction. Environmental Science & Technology. 2006;**40**(17):5225-5232

[71] Evangelou MW, Ebel M, Schaeffer A. Chelate assisted phytoextraction of potentially-toxic elements from soil, effect, mechanism, toxicity, and fate of chelating agents. Chemosphere. 2007;**68**(6):989-1003

[72] López-Vizcaíno R, Alonso J, Cañizares P, León MJ, Navarro V, Rodrigo MA, et al. Electroremediation of a natural soil polluted with phenanthrene in a pilot plant. Journal of Hazardous Materials. 2014;**265**:142-150

[73] Paz-García JM, Johannesson B, Ottosen LM, Alshawabkeh AN, Ribeiro AB, Rodríguez-Maroto JM. Modeling of electrokinetic desalination of bricks. Electrochimica Acta. 2012;**86**:213-222

[74] Maturi K, Reddy KR, Cameselle C. Surfactant-enhanced electrokinetic remediation of mixed contamination in low permeability soil. Separation Science and Technology. 2009;**44**(10):2385-2409

[75] Reddy KR, Maturi K, Cameselle C. Sequential electrokinetic remediation of mixed contaminants in low permeability soils. Journal of Environmental Engineering. 2009;**135**(10):1943-7870

[76] Reddy KR. Electrokinetic remediation of soils at complex contaminated sites: Technology status, challenges, and opportunities. In: Manassero M, Dominijanni A, Foti S, Musso G, editors. Coupled Phenomena in Environmental Geotechnics. London, UK: CRC Press; 2013. pp. 131-147

[77] Bahemmat M, Farahbakhsh M. Catholyte-conditioning enhanced electrokinetic remediation of Co and Pb polluted soil. Environmental Engineering and Management Journal. 2015;**14**:89-96